New Challenges New Tools for Defense Decisionmaking

Stuart E. Johnson   Martin C. Libicki   Gregory F. Treverton

Bruce W. Bennett   Nurith Berstein   Frank Camm
David S.C. Chu   Paul K. Davis   Daniel B. Fox   James R. Hosek
David Mussington   Stuart H. Starr   Harry J. Thie

RAND

2003

This publication was supported by RAND using its own funds.

Library of Congress Cataloging-in-Publication Data

New challenges, new tools for defense decisionmaking / edited by Stuart Johnson,
Martin Libicki, Gregory F. Treverton.
    p. cm.
    "MR-1576."
    Includes bibliographical references.
    ISBN 0-8330-3292-5 — ISBN 0-8330-3289-5 (pbk.)
    1. United States—Military policy—Decision making. 2. National security—
United States. 3. United States—Defenses. 4. World politics—21st century. I.
Johnson, Stuart E., 1944– II. Libicki, Martin C. III. Treverton, Gregory F.

UA23 .N374 2003
355'.033573—dc21

                                                                    2002190880

Cover design by Peter Soriano

Published 2003 by RAND
1700 Main Street, P.O. Box 2138, Santa Monica, CA 90407-2138
1200 South Hayes Street, Arlington, VA 22202-5050
201 North Craig Street, Suite 202, Pittsburgh, PA 15213-1516
RAND URL: http://www.rand.org/
To order RAND documents or to obtain additional information,
contact Distribution Services: Telephone: (310) 451-7002;
Fax: (310) 451-6915; Email: order@rand.org

This book contains thirteen papers that, collectively, provide a wide variety of perspectives on future defense decisionmaking. The topics range from the global security environment, to tools for assessing both military manpower and information systems, to general techniques for gauging the adequacy of military forces, to specific techniques for improving defense decisionmaking, especially under conditions of uncertainty. The papers were originally delivered as lectures in a short-course series, "Defense Analysis for the 21st Century," that was sponsored by the RAND Graduate School and held in RAND's Washington, DC, offices.

The rethinking of defense planning will continue apace, spurred in particular by September 11, 2001, and the war against terrorism; but many of the ideas and conclusions presented here will retain their relevance for years to come. While no attempt has been made to develop a monolithic "RAND view" on the many challenges defense planners face, readers will see in this collection themes and methodologies characteristic of RAND's recent defense planning work. It is hoped that these will be of interest to a broad class of individuals in government, military service, universities, and industry. The intention has been to provide papers that both describe enough of the research and analytic reasoning to convey a sense of how the work was done and present findings that should be useful for some time to come.

Most of the research underlying the papers in this book was conducted in RAND's three national security federally funded research and development centers (FFRDCs): Project AIR FORCE, the Arroyo

Center, and the National Defense Research Institute (NDRI). These three centers are sponsored, respectively, by the Air Force, the Army, and the Office of the Secretary of Defense and the Joint Staff.

RAND's National Security Research Division oversaw the preparation of this book for distribution within the defense analysis community. This activity was made possible by funding from RAND's continuing program of self-sponsored independent research, which is derived, in part, from the independent research and development provisions of RAND's contracts for the operation of its three U.S. Department of Defense FFRDCs.

CONTENTS

Part III. New Tools for Defense Decisionmaking

Chapter Nine
EXPLORATORY ANALYSIS AND IMPLICATIONS FOR
MODELING

Chapter Ten
USING EXPLORATORY MODELING

FIGURES

TABLES

ACKNOWLEDGMENTS

Only editors can know how much help it takes to produce a book like this. The authors, first and foremost, suffered, more or less patiently, through delays and revisions, some of them extensive, in the interest of converting a clutch of good ideas into a relatively coherent volume. Bob Klitgaard, dean of the RAND Graduate School, had the idea for the book and was its sponsor throughout. We received, in RAND fashion, reviews that were both detailed and frank from Ted Harshberger, Tom McNaugher, and Michael O'Hanlon. One of us, Greg Treverton, started as a reviewer but was then challenged to put a contribution where his critique had been. He met the challenge.

We had the good fortune to have Jeri O'Donnell as editor and Janet DeLand doing the final page composition. If important analytic concepts are seen to emerge from complicated prose, that is a tribute to their fine work. Last, but hardly least, Rachel Swanger shepherded the book to final completion with grace and persistence.

To all these good people, we express our appreciation. We also repeat the usual disclaimer that while they can take credit for what is good in these pages, only we and the authors should be held responsible for any gremlins that remain.

ABBREVIATIONS

A&T	Acquisition and Technology
ABAT	Army brilliant antitank
ACAV	air-cushioned amphibious vessel
AFQT	Armed Forces Qualification Test
APOD	airport of debarkation
ASD	Assistant Secretary of Defense
ASVAB	Armed Services Vocational Aptitude Battery
ATACM	Army's Tactical Missile System
AWACS	Airborne Warning and Control System
BAT	brilliant antitank
BCP	best commercial practice
bps	bits per second
BRAC	base realignment and closure
BW	biological weapons
CAIV	cost as an independent variable
CAPE	C4ISR Analytic Performance Evaluation
CBW	chemical and biological weapons
CCD	charge-coupled device
C-Day	day on which U.S. forces begin mobilizing
CD-ROM	compact disk—read-only memory
CEC	cooperative engagement capability
CFATF	Caribbean Financial Action Task Force

CFE	Conventional Forces in Europe
C4	command, control, communications, and computers
C4ISR	command, control, communications, computers, intelligence, surveillance, and reconnaissance
CICA	Competition in Contracting Act
CINC	commander in chief
CMOC	civil-military operations center
COBP	code of best practice
CONOP	concept of operations
CONUS	continental United States
CPO	chief purchasing officer
C3	command, control, and communications
C3I	command, control, communications, and intelligence
C2	command and control
CVBG	carrier battle group
CW	chemical weapons
DAB	Defense Acquisition Board
DCAA	Defense Contract Administration Agency
D-Day	day on which war begins
DEPTEMPO	tempo of deployments
DMSO	Defense Modeling and Simulation Organization
DoD	Department of Defense
DoN	Department of the Navy
DOS	disk operating system
DOTML-PF	doctrine, organization, training, materiel, leadership and education, personnel, and facilities
DP	dimensional parameter
DRB	Defense Resources Board
DRID	Defense Reform Initiative Directive
DSD	Decision Support Department
DVD	digital versatile disk

EELV	Evolved Expendable Launch Vehicle (U.S. Air Force)
EMS	electronic meeting system
EPCRA	Emergency Planning and Community Right-to-Know Act
EU	European Union
FAIR	Federal Activities Inventory Reform
FAR	Federal Acquisition Regulations
FCS	Future Combat System
FEMA	Federal Emergency Management Agency
FinCEN	Financial Crimes Enforcement Network
FY	fiscal year
FYDP	Future Years Defense Program
GCC	Gulf Cooperation Council
GCCS	Global Command and Control System
GCSS	Global Combat Support System
GIG	global information grid
GNP	gross national product
GPS	global positioning system
HLA	High Level Architecture
HPM	high-power microwave
HTML	hypertext markup language
HUMINT	human intelligence
ICBM	intercontinental ballistic missile
IFFN	Identification Friend Foe or Neutral
IPT	integrated process team
IRA	Irish Republican Army
ISO	International Standards Organization
ISR	intelligence, surveillance, and reconnaissance
IW	information warfare
JCS	joint chiefs of staff
JDAM	Joint Direct Attack Munition
JFCOM	Joint Forces Command

JICM	Joint Integrated Contingency Model
JPEG	Joint Photographic Experts Group
JSF	Joint Strike Fighter
JSIMS	Joint Simulation System
JSOW	Joint Standoff Weapon
JSTARS	Joint Surveillance Target Attack Radar System
JVB	Joint Virtual Battlespace
JWARS	Joint Warfare System
km	kilometer
Kt	Kiloton
LAN	local area network
M&S	modeling and simulation
MCES	Modular Command and Control Evaluation Structure
MEMS	micro-electromechanical system
MIT	Massachusetts Institute of Technology
MLRS	multiple-rocket launcher system
MOA	mission oriented approach
MOE	measure of effectiveness
MOFE	measure of force effectiveness
MOM	measure of merit
MOOTW	military operations other than war
MOP	measure of C2 system performance
MORS	Military Operations Research Society
MRM	multiresolution modeling
MRMPM	multiresolution, multiperspective modeling
MTW	major theater war
NATO	North Atlantic Treaty Organization
NBAT	Naval brilliant antitank
NCO	noncommissioned officer
NGO	nongovernmental organization
NIPRnet	joint Internet for unclassified work

NPS	Naval Postgraduate School
NRL	Naval Research Laboratory
OFW	Objective Force Warrior
OMB	Office of Management and Budget
ONR	Office of Naval Research
OOTW	operations other than war
OPM	Office of Personnel Management
OSD	Office of the Secretary of Defense
PEM	personal-computer model
PGM	precision guided munition
PO	petty officer
POM	program objectives memorandum
PPBS	Planning, Programming, and Budgeting System
PSYOP	psychological operation
QDR	Quadrennial Defense Review
RAM	random access memory
R&D	research and development
RMA	revolution in military affairs
ROTC	Reserve Officers Training Corps
RSTA	reconnaissance, surveillance, targeting, and acquisition
SADARM	sense and destroy armor munition
SALT	Strategic Arms Limitations Treaty
SAM	surface-to-air-missile
SARA	Superfund Amendments and Reauthorization Act
SAS	Studies, Analysis, and Simulations
SBA	Small Business Administration
SBCT	Stryker Brigade Combat Team
SCAV	Sea Cavalry
SEAD	suppression of enemy air defenses
SFW	sensor fused weapon
SINCGARS	Single Channel Ground and Airborne Radio System

SIPRnet	joint Internet for secret material
SIW	strategic information warfare
SLBM	submarine-launched ballistic missile
SOF	special operations force
SPOD	seaport of debarkation
SROC	Senior Readiness Oversight Council
SSC	smaller-scale contingency
SWA	Southwest Asia
TACCFS	Theater Air Command and Control Simulation Facility
TBMD	theater ballistic missile defense
TCO	transcontinental criminal organization
TCP/IP	transmission control protocol/Internet protocol
TDA	table of distribution and allowance
TOC	tactical operations center
TOC	total ownership cost
TOE	table of equipment
TQM	total quality management
TRI	toxic releases inventory
UAV	unmanned aerial vehicle
UCAV	unmanned combat air vehicle
VV&C	verify, validate, and certify
WMD	weapons of mass destruction
XML	extensible markup language

INTRODUCTION

Stuart E. Johnson, Martin C. Libicki, and Gregory F. Treverton

It is commonplace to say, but still easy to underestimate, how much the collapse of the Soviet Union and the end of the Cold War transformed the task of U.S. foreign and defense policymaking. And the terrorist attacks of September 11, 2001, have opened yet another era, one whose shape and dimensions are yet to be understood. This volume addresses the new challenges of this changed world, the difficulties for defense planning that these challenges engender, and new analytic techniques that have been developed at RAND and elsewhere for framing particular problems.

During the Cold War, the Soviet threat provided the benchmark for both foreign policy and defense: containing that threat was the overriding purpose of policy, and the threat provided the organizing principle for shaping U.S. military forces. If U.S. forces could deter or, if need be, defeat the Soviet military, they could handle other threats as "lesser included cases."

In this context, defense planning was dominated by "force-on-force" comparisons of U.S. and Soviet forces. While the United States spent billions of dollars trying to learn the specific numbers and capabilities of Soviet weapons, overall the Soviet threat changed only gradually and fairly predictably. What Moscow would have tomorrow could generally be predicted to be bigger and slightly better than what it had today. The Soviet defense establishment might float on a sea of shoddiness in the rest of the Soviet economy, but the Soviets could be presumed willing and able to spend what it took to keep pace with the United States.

1

All of this changed with the end of the Cold War. In place of (albeit sometimes terrifying) predictability, the world became very unpredictable. In place of a single overriding threat that could be used as a benchmark for measuring everything else, a number of possible threats arose, some of them potentially very dangerous if not, ultimately, in a league with the Soviet threat. In place of the threat of force-on-force engagements with a strong foe, there were now "asymmetric" threats from potential U.S. foes much weaker than the United States. Such foes would not be foolish enough to take America on directly (as Saddam Hussein did), instead pursuing the ancient art of strategy long obscured by a bipolar U.S.-Soviet competition. They would look for U.S. vulnerabilities, ways not to defeat U.S. power but to render it irrelevant.

In one respect, defense decisionmaking has not changed. From hot war to cold, to a time of terrorism, the U.S. Department of Defense (DoD) has remained one of the world's largest and most complex organizations. It employs upward of 2.4 million people (military and civilian) and contracts for the direct services of another 400,000. It runs a budget of $400 billion and an accumulated capital stock (real property and equipment) of more than $4 trillion. Managing an organization of this magnitude is a daunting job. Decisions affect multibillion-dollar programs, tens of thousands of people, and, most important, the security of the nation itself.

FROM OLD CHALLENGES TO NEW

During the long Cold War, from the 1950s into the 1980s, RAND, often working with other analytic organizations, played a key role in developing techniques to inform decisionmaking and guide the development of national strategy. Notable RAND contributions to techniques for defense decisionmaking include

- The marrying of blast physics, ballistics, and guidance technology to determine "how much is enough" to hold the Soviet economy and population at risk so as to deter the Soviets from launching a nuclear first strike.

- The application of game theory to determine the appropriate size and configuration of U.S. strategic nuclear forces.

- The application of "armored division equivalents" to clarify the strengths and weaknesses of NATO strategy to defend Western Europe.

- The development of the strategy-to-task methodology to provide a framework for weapon systems evaluation.

These and other techniques were disseminated widely and gained broad acceptance within the defense analysis community. They provided a common framework in which debates over budget allocations, investment strategy, and doctrine development could take place. Although far from perfect in predicting future requirements, these techniques, when broadly shared, managed to carry the debate beyond "strongly held opinion." Positions now had to withstand the test of objective, systematic, and (where possible) quantitative analysis.

The practice of using analytic techniques to illuminate the options available and to provide a framework for such far-reaching decisions began in earnest with Robert McNamara, who served as U.S. secretary of defense from 1961 to 1967 under Presidents John Kennedy and Lyndon Johnson. McNamara brought together a cadre of analysts, led by several RAND researchers, to establish the Office of Systems Analysis as the focal point for changing the basis of DoD decisionmaking from "strongly held views" to objective, thorough analysis. This office had, as all new institutions do, its ups and downs, its supporters and detractors. But as it began to influence important decisions—affecting, for example, the number of U.S. intercontinental ballistic missiles (ICBMs) or the size and disposition of U.S. forces committed to the North Atlantic Treaty Organization (NATO)—the value of an analytic basis for debating decisions became clear. Before long, the Air Force and then other military services and the joint chiefs of staff (JCS) established their own enhanced analytic capability.

In the 1980s, as attention turned to serious arms control negotiations, analytic techniques were adapted to provide a strategic framework for U.S. negotiators in, successively, the Strategic Arms Limitations Talks (SALT) and Strategic Arms Reduction Treaty (START) negotiations over strategic nuclear forces. Similarly, when the United States and the Soviet Union, along with their respective allies, began to discuss in earnest the size of conventional forces in Europe, U.S. negotiators were well informed about which Warsaw Pact forces posed the greatest threat and which NATO forces were most critical to maintaining a stalwart defense of NATO territory. The resulting treaties strengthened the U.S. position and enhanced strategic and conventional stability. Indeed, the START II and Conventional Forces in Europe (CFE) treaties remain, even today, important tools in U.S. national security policy.

These analytic techniques were developed for the specified Cold War requirement: face a large military capable of challenging the United States across the globe and throughout the escalation ladder—from local conflicts involving proxy forces to a strategic nuclear exchange. In this context, the techniques led to measures of military effectiveness (such as armored division equivalents), simulations of large conventional force-on-force engagements, probabilities of killing a missile silo, and simulations of a nuclear exchange.

After Moscow abandoned superpower competition with the United States and the Cold War ended, the requirement that dominated U.S. defense planning changed: to prevail in two nearly simultaneous major theater wars (MTWs). Planning documents typically cited aggression by Iraq or North Korea against its neighbors, attacks that have been the subject of considerable analysis, modeling and simulation, intelligence research, and war gaming. Because such operations would entail large formations of classical military forces, analyses of them have drawn heavily on techniques developed during the Cold War. But even this application of Cold War analytic techniques is losing its utility. Iraq's armed forces suffered from obsolescence as a decade of economic sanctions and lack of access to modern military equipment took its toll. The armed forces of North Korea fared even worse.

As it is, the global security environment is profoundly changed. Preoccupation with the daunting though well-ordered Cold War threat has been replaced by the far different set of challenges cited by recent defense secretaries, challenges driven home by the tragedy of September 11:

- Countering terrorism
- Countering the spread of weapons of mass destruction (WMD)

- Peace enforcement

- Crisis response

- Enforcement of economic and military sanctions

- Combating narcotics trafficking

The task of meeting these challenges is critical to maintaining the economic prosperity and free exercise of democratic governance valued by the United States and its allies. This task is not advanced by naïvely applying analytic and planning techniques developed and refined in the Cold War and appropriate to dealing only with large-scale, cross-border aggression. Planners require different tools. To be sure, even conventional adversaries can inflict serious damage on U.S. and allied forces and interests, but adversaries are likely to turn to other means of challenging us—employing WMD or terrorism. This poses a dilemma for decisionmakers: the analytic tools at their disposal credibly address a problem of declining probability, even as they do little to illuminate the challenges U.S. forces are increasingly being called on to respond to.

During the dozen years since the dissolution of the Soviet Union, the research staff at RAND has worked intensively to adapt traditional defense analysis techniques to today's security environment and to develop new techniques as necessary. These techniques address today's key questions:

- How can the United States set requirements in the face of an uncertain future?

- How can the nation plan for flexibility and program it into our forces?

- How might rapid advances in commercial technologies help the armed forces?

- How has the profile of skills needed by the armed forces changed? And what needs to be done to recruit, train, and retain the required cadre of skilled personnel?

Taken together, the chapters in this volume provide a new portfolio of tools to frame decisions, to solve problems, and to analyze alternatives.

HOW THIS VOLUME IS ORGANIZED

This book comprises thirteen chapters organized into three parts:

I. New Challenges for Defense

II. Coping with Uncertainty

III. New Tools for Defense Decisionmaking

Part I begins with a chapter describing the questions that any nation must answer about its defense—how much, what, and how?—and provides an overview of the structures, especially inside the Pentagon, that the United States has developed to answer them. Chapter Two assesses the challenge of chemical and biological weapons as one class of "asymmetric" strategies a U.S. adversary might employ. The logic of that approach translates to terrorism, in many respects the ultimate asymmetric strategy of the weak against the strong. The subject of the third chapter is the "information architecture" needed for U.S. defense. The rise of the information age coincided with, indeed probably hastened, the collapse of the Soviet Union, and the United States now confronts the challenge of how to structure itself to best take advantage of the continuing revolution in information technology.

Part II of this book takes up the driving challenge of uncertainty. The first chapter is based on the premise that while the future is unknowable, it is not a complete mystery. Broad trends are discernible, one being the continuous improvement in information technology, which suggests that future warfare will be information based, with huge amounts of information from various sensors integrated to build a picture of the battlespace. The second chapter here describes "uncertainty-sensitive planning." Cold War planning may have been based on threats, but the future cannot be, for there is simply too much uncertainty, too many diverse threats. Planning needs to aim at building a portfolio of capabilities designed not just to confront future threats but also to hedge against them and to shape the environment so they do not develop.

The next two chapters in Part II are on planning the human resources for tomorrow's defense. RAND research was the analytic basis for the U.S. shift from conscription to an all-volunteer force in the 1970s and has contributed to the great success that force has been. These chapters outline visions of the future soldier, specify the factors critical to attracting and retaining the talent needed for a volunteer force, and describe a planning mechanism for shaping the force. The final chapter looks at the challenge of adopting best commercial practices from the business world into the apparently similar but actually quite different world of defense planning and management.

Part III begins with a chapter on exploratory modeling, a tool growing out of new techniques in modeling and the computer power that now exists. The technique fits well with capabilities-based planning in that it permits analysts to move far beyond a few canonical scenarios to examine a wide array of variables, looking for key uncertainties and key drivers across a wide range of possibilities. The second chapter in Part III is a more concrete illustration of how exploratory modeling can be employed. The third chapter chronicles the long effort to assess how much information systems contribute to effectiveness in conducting military missions—a quest, needless to say, that has been reshaped by the Cold War's demise and the diversity of future missions.

The penultimate chapter's subject is the "Day After" gaming technique that RAND developed to examine strategic issues. Essentially, this technique entails playing out a scenario and then working backward to see how better decisions at an earlier stage might have improved the outcome by mitigating threats or providing more options. The chapter explains and illustrates the technique. The final chapter then describes another tool, electronic meeting systems, and elaborates with an example of its use—the U.S. Navy preparing for the 1997 Quadrennial Defense Review—that led to interesting, perhaps unconventional results.

An Afterword concludes the book. It suggests some of the implications of the current war on terrorism and homeland security for the challenges that lie ahead. These challenges will require RAND and its fellow organizations to reshape, once again, analytic approaches that improve defense and the national security decisionmaking.

PART I. NEW CHALLENGES FOR DEFENSE

INTRODUCTION TO PART I

Defense planning during the Cold War was dominated by the threat from the Soviet Union. It was, in that sense, threat based. It also was, to a great extent, symmetrical, based on force-on-force calculations for U.S. and Soviet armored forces, fighter jets, and the like. In these circumstances, the U.S. planning structure within the Pentagon became increasingly centralized, seeking to maximize the benefits from various investments in ways to better cope with the Soviet threat.

All the practices that made considerable sense during the Cold War badly need to be rethought now. Soviet strategy may have been more creative than it was usually given credit for, but it was relatively slow moving. By contrast, today's threats—and still more tomorrow's— are many and very uncertain. While none may be in a class with the Soviet threat, the attacks of September 11, 2001, drove home how lethal even "lesser" threats can be. Moreover, U.S. military power has given rise to a paradox: the United States is so dominant in its ability to fight a conventional armored war that it is not likely to have to fight such a war. Realizing the futility of a conventional face-off with the United States, would-be adversaries will instead aim to confront the United States where it is weak or can be surprised—posing what are called *asymmetric* threats. Terrorism, the strategy of the weak against the strong, is quintessentially an asymmetric strategy.

This change from a fairly predictable, symmetrical threat to the myriad unpredictable, asymmetrical threats possible has profound effects for defense planning. It impels a shift from threat-based planning to capabilities-based planning and suggests that a "portfolio" approach to these capabilities—i.e., trying to build breadth and flexibility in the hope that capabilities can be brought to bear across a spectrum of unpredictable threats—would be the most useful type. It also presses the United States to draw on advantages where it has them, particularly in harnessing information technology to identify threats, link shooters tightly to sensors, and manage a flexible, fast-moving campaign. And it hints at the value of decentralizing Pentagon management so as to encourage the innovation needed to produce a real "revolution in military affairs" (RMA).

In the first chapter here, "Decisionmaking for Defense," David S.C. Chu and Nurith Berstein argue that any military organization must ask itself four questions: what forces should be fielded, how should they be trained, how should they be equipped, and what tempo of operations should they maintain? The two questions on force structure and equipment are likely to dominate in the next decade. The debates over the size and structure of military forces that were not fully joined in the 1990s will have to be faced squarely, particularly in light of strains placed on the current force by the pace of operations, all the more so with the war on terrorism. Moreover, the urgent need to recapitalize the present generation of equipment places this issue near, perhaps atop, the agenda. In this regard, a central issue is the degree to which new investment should shift from modernizing existing capabilities to procuring quite different capabilities, ones geared to a different vision of what future military forces might look like. The more decisionmakers lean toward a new vision, the greater the challenge they will pose to how "legacy" systems—including some still under development—are treated. Such fundamental issues are never settled once and for all or even for very long.

In Chapter Two, "Responding to Asymmetric Threats," Bruce Bennett explains why U.S. conventional military superiority has forced adversaries to pursue asymmetric strategies—i.e., those designed to attack such vulnerabilities as U.S. and allied will, host nation support, and basing infrastructure. Potential adversaries have developed weapons of mass destruction (WMD), information warfare, and simple countermeasures such as sea mines as part of their asymmetric threats, and they use camouflage, concealment, and deception to hide their capabilities and strategies. Part of the danger of asymmetric threats stems from the surprise they can achieve, which undercuts U.S. response preparation, leaving the United States at a disadvantage. Asymmetric threats can affect U.S. and allied forces, civilians, and interests in diverse ways that are difficult and expensive to counter. The threat of retaliation, alone, is insufficient to deter their use in at least some cases. What is needed is an integrated defense effort that includes three elements: understanding the threats, protecting against them, and threat management. Understanding the threats is key to addressing them, protection is necessary to reduce any gains the adversary may be seeking, and threat management seeks to deter the spread of such threats and discourage the development of new ones. The United States must institutionalize its response efforts within its own military; it must also internationalize them, coordinating with allies to provide a common defense.

Chapter Three, "What Information Architecture for Defense?" by Martin Libicki, is a plea for planners to recognize the choices, deliberate or not, that underlie enterprise-level information technology systems and thereby shape how they are used. One office networking system may be much like another, but the Pentagon's requirements for information will vary depending on whether it is planning for a strategic campaign (putting a premium on analytic skills to determine enemy strengths and weaknesses), a conventional campaign (with its need for mass force coordination), modern high-technology warfare (which requires that targets be found and prosecuted in real time), or a low-intensity conflict (which requires that warfighters be enabled with subtle but detailed portraits of their environment). Information architectures may be described by how they collect, present, display, circulate, maintain, secure, standardize, and integrate information. Each of these eight dimensions involves choices that the Pentagon must make if it is to make best use of U.S. advantages in information technology.

Chapter One

DECISIONMAKING FOR DEFENSE

David S.C. Chu and Nurith Berstein

Defense is, for all nations, at the heart of national security. All nations face a common set of choices—what decisions must be made, who will make them, how resources will be allocated, and what investments will be made. At one level up, nations have to decide what principles and style of decisionmaking are appropriate for them, and, importantly, what structure will govern the process of defense decisionmaking. This chapter discusses these choices and reviews the issues that must be addressed in devising a governance structure for making them, drawing on U.S. experiences over the last half century. It concludes with a short discussion of alternative approaches and styles before looking briefly to the future.

America's experiences may have lessons for others even if their circumstances dictate a different set of governance arrangements for defense decisionmaking. Equally important, the United States is now at a point in its history when it must reconsider—if only to reconfirm—its own governance structure. The Cold War that motivated so much of the U.S. defense establishment and shaped its decisionmaking mechanisms has been replaced with a much different set of security challenges. The technological assumptions on which so many of DoD's current choices rest also must be reconsidered. Thoughtful defense analysts argue that a "revolution in military affairs" (RMA) is, and should be, under way. In short, should the United States in the early 21st century continue to make defense decisions the way it did in the latter half of the 20th?

DECIDING WHAT DECISIONS MUST BE MADE

Every defense establishment faces a set of interrelated decisions that it must make and that its governance structure should be designed to confront:

- What set of forces should the country maintain? How should forces be organized? Under what command structure?

- What training should forces receive? How ready should they be, and for what?

- With what equipment should forces be armed? In what condition should equipment be maintained?

- What tempo of operation should forces be prepared to maintain? What stock of consumable items and spare parts should be stockpiled to support this tempo? What ongoing maintenance capability is needed to sustain this pace of operation?

These decisions govern what the defense establishment delivers, but they should be guided by the *outcomes* desired by the national leadership. For the past 25 years, DoD has translated these outcomes into scenarios against which U.S. military forces are measured. During the late Cold War, the planning scenario focused on global conflict with the Soviet Union (on two fronts, Europe and Southwest Asia). After the Cold War, this scenario was replaced by a requirement to conduct two nearly simultaneous major theater wars (MTWs) while also conducting operations other than war (e.g., peacekeeping in the Balkans). When pressed for specificity, DoD posited the two MTWs as being on the Korean peninsula and in Southwest Asia.

When the Cold War ended, DoD tried at one point to formulate a new structure in which to make decisions about U.S. military forces—forces would be judged not against specific scenarios but against a set of military capabilities the country should maintain. DoD wanted to move away from a single scenario; its military leadership was concerned that no single scenario would be compelling. The shortcoming of the capabilities approach, as articulated in testimony by then Secretary of Defense Les Aspin, was that it did not yield defensible, specific criteria against which to judge military forces. To define such specific criteria, DoD tried generic "illustrative planning scenarios." The lack of geographic specificity in these scenarios, however, when applied in the debate over the acquisition of the C-17, proved their undoing. DoD reverted to the concrete illustrations of conflict in the Persian Gulf and Korea, from which the notion of two nearly simultaneous MTWs eventually developed.[1]

DECIDING WHO MAKES THE DECISIONS

A notable feature of the American political landscape is the U.S. Congress's salient role in defense decisionmaking, which is spelled out plainly in the Constitution. In enumerating the powers of the Congress, Article I gives it the authority to declare war, to raise and support armies and provide and maintain a navy, and to establish rules for the governance of the military. Indeed, of the 18 congressional powers enumerated in Section 8 of Article I, five explicitly deal with the military.[2]

The creation of a Secretary of Defense in 1947 reflected a balance between the prerogatives of the individual military services and President Truman's desire for a central executive to coordinate and rationalize their separate activities. The first secretary, James Forrestal, resigned after a largely unsuccessful struggle to orchestrate the activities of the National Military Establishment (as it was then called), frustrated by his limited powers as secretary. The 1949 amendment of the National Security Act addressed some of these limitations. It created the Department of Defense, subordinated the military departments to the secretary, and strengthened the staff supporting the secretary. Amendments enacted in 1958 further enhanced the secretary's role, thus paving the way for the far-reaching changes Robert McNamara imposed on the department. But DoD governance retains a tension between the centrifugal, competitive forces reflected in the responsibilities of the individual military departments (in whose well-being Congress takes a deep interest) and the centralizing responsibilities of the defense secretary.

---

[1]Early in the Clinton administration, DoD leadership considered a posture of preparing for one MTW while checking a second opponent until resources could be mobilized or released to deal with it ("win—hold—win"). The resulting political uproar convinced the administration to endorse the two-MTW standard.

[2]The important (but often neglected) role of Congress is discussed usefully in Charles A. Stevenson, "Bridging the Gap Between Warriors and Politicians," paper for the 1999 Annual Meetings of the American Political Science Association, Atlanta, GA.

There is a further division of authority within the military departments, the one between civilian political appointees and the uniformed military hierarchy. This split is reflected in the fact that the separate civilian secretariat reports to the secretary of the military department, whereas the uniformed staff reports to the chief of staff. Much of the statutory authority wielded by a military department is actually held by that department's secretary, even though the uniformed staff is much larger than the civilian secretariat and typically exercises de facto control of the day-to-day agenda.

The Goldwater Nichols Act of 1986 changed the division of defense authority in three important ways. First, within the military departments, it strengthened the hand of the civilian service secretariats by formally subordinating the uniformed officers previously responsible for weapons acquisition and budget execution to their civilian counterparts rather than to the service chief of staff. For several decades, acquisition authority in the military departments had been divided between a civilian assistant secretary and a military deputy chief of staff assigned that function. Likewise, each military department had a military comptroller who reported through the chief of staff rather than to the civilian counterpart in the Office of the Secretary of Defense (OSD) responsible for financial matters. The Goldwater Nichols Act required that these military officers report to the civilian counterpart.

Second, Goldwater Nichols ratified the expanded authority of the commanders in chief of the unified and specified commands (CINCs), who, to the discomfit of the military departments, had been invited by Secretary Caspar Weinberger in 1981 to play a significant role in DoD's resource allocation processes. Goldwater Nichols further reinforced the CINCs' authority by requiring that all military units be assigned to one of their commands. Moreover, the CINCs were explicitly made responsible for the preparedness of their commands to carry out assigned missions. These changes solidified the CINCs' role as "customers" of DoD and, especially, of the military departments. The Act also underscored the future importance of joint operations as the way U.S. forces would be employed in the field, and thus the way in which planning for them should be conducted, including planning undertaken by the military departments. One example of this increased emphasis on "jointness" is that the annual DoD budget proposals submitted to Congress include a separate item for joint exercises undertaken by the commands.

Third, Goldwater Nichols further empowered the chairman of the joint chiefs of staff (JCS) and joint staff. In the years following World War II, that position had gradually evolved into one clearly seen as the nation's senior military officer. Although the chairman is not legally part of the chain of command—which runs directly from the president through the secretary of defense to the CINCs—his advice is often treated with the same deference as that of the defense secretary, especially by Congress. In these ways, the Goldwater Nichols Act strengthened the chairman's advisory role, causing considerable concern within the military departments that his responsibilities importantly infringe on what they believe should be their responsibilities.

The Act also produced, in combination with the distinctive events of the last 15 years, a new central actor, the joint staff, which is in tension with the military departments because of its perceived intrusion on their authority (reminiscent of that produced by the "whiz kids" of Secretary McNamara's staff in the 1960s).

Divided authority could be a formula for bureaucratic gridlock and inaction, with many having the right to say "no," but no element strong enough to see a program proposal through to approval and successful execution. One of the mechanisms that DoD has used in this circumstance, both to secure a wide circle of advice and to forge consensus on the best course of action, is the advisory board—i.e., a formal body that gives many if not all parties a "voice" in the process while allowing final decisionmaking authority to remain in the hands of the board's chair. The most powerful senior-level boards are

- The Defense Resources Board (DRB), chaired by the deputy secretary of defense. Advises the deputy secretary on major resource allocation decisions.

- The Defense Acquisition Board (DAB), chaired by the under secretary of defense for acquisition and technology (A&T). Advises the under secretary (A&T) on major acquisition programs and acquisition policies and procedures.

- The Joint Requirements Oversight Council, chaired by the vice chairman of the joint chiefs of staff (who also serves as the vice chairman of the DAB). Validates mission needs developed by the CINCs and by planning elements of the joint staff, reviews performance parameters and requirements, and develops recommended joint priorities for those needs.

- The Senior Readiness Oversight Council (SROC), chaired by the deputy secretary of defense. Advises the secretary of defense on readiness, oversees readiness-related actions, reports on relevant readiness questions, and coordinates DoD positions on readiness for outside audiences.

Each board was created by the direction of, or with support from, a particular secretary of defense, although succeeding secretaries have used and shaped them in accord with their styles. Thus, while the formal roles of these boards often change little over time, their real roles and authority respond to the style of each secretary, giving each secretary considerable latitude in how the department is managed.

Notably absent from this description of who makes decisions on defense issues is the U.S. president and his immediate staff. Designated by the Constitution as the commander-in-chief, the president could, in principle, take a detailed role in defense decisionmaking. The president and his staff typically do take an active role in formulating national security strategy, thus setting the basic course for the defense establishment, and the president usually makes the key operational decisions in times of crisis. But, otherwise, the American practice has been to leave most department managerial decisions to the defense secretary, although the president does set the budgetary constraint within which the department must live.

In the Kennedy administration, a concerted effort was made to involve the president early in key defense decisions. It was felt that securing the president's guidance early in the decisionmaking cycle would help the department formulate better policies. Draft presidential memoranda were prepared as vehicles for raising issues with the president. But when the first of these was presented to President Kennedy, he indicated that he was not prepared to make choices so early. The memoranda lived on for a period as a useful way to conduct policy debates within DoD, but they were never more than drafts and were never again sent to the president.

DECIDING HOW TO ALLOCATE RESOURCES

Budgets in bureaucracies are typically created one year at a time and are based disproportionately on expenditure patterns of the prior year. A group of analysts at RAND in the 1950s developed an alternative approach to budget preparation, one based on the idea that the proper way to begin was by setting long-term objectives. Codified under the cumbersome title Planning, Programming, and Budgeting System (PPBS), Robert McNamara brought the ideas behind this approach to the Pentagon in the 1960s when he hired Charles Hitch as comptroller from his prior post as head of RAND's Economics Department.[3]

The planning phase of the PPBS sets long-term goals. The secretary of defense announces objectives for the department in what is now called the Defense Planning Guidance. The Guidance is ultimately the secretary's document, although his own staff, the military departments, and the chairman and his staff all participate, reflecting the multiple centers of authority within the department. The document includes a variety of ways to measure progress toward the secretary's goals, including a set of illustrative scenarios describing the military events the secretary believes should guide key decisions of the department.

As administered since the late 1960s, the programming stage of the PPBS consists of the three military departments preparing a set of fiscally constrained proposals to meet the secretary's goals. These program objectives memoranda, or POMs, extend six years into the future. The secretary's office reviews the POMs to ensure they conform with the guidance provided by the secretary in the planning phase. Changes are made as required. Although the programming phase is a debate about means—which program choices best achieve the stated goals—it often reopens the debate about those goals, revisiting choices made in the planning phase.

---

[3]The spirit of the Government Performance and Results Act distinctly parallels that of the PPBS.

Once decisions about the six-year program are made, the material in the POMs is consolidated into the Future Years Defense Program (FYDP), and the department is then ready to formulate its budget for the next fiscal period. The department's constituent elements prepare budget estimate submissions based on the program decisions, reflecting latest pricing and execution experience. These are reviewed by the secretary's office, in a joint process with the Office of Management and Budget (OMB), and consolidated in the president's budget request.

The sharing of authority in PPBS reflects the reality of DoD's divided authority. It gives each element of DoD (most especially the military departments) a chance to fashion its future course within the parameters set by the secretary of defense and subject to his review and final decision. But the parameters are debated with the many elements before they are set, and the reviews of both the program and the budget include the affected parties, which are allowed wide latitude to argue their cases before the secretary makes final decisions.

Nonetheless, PPBS gives the secretary of defense the essential tool to control the department's key decisions, each of which requires resources to implement: the structure of forces, their training and readiness, the equipment with which they are armed, and the provisions set aside to sustain them in operations. At the same time, both the 1969 decision to give each of the department's constituent elements the right to prepare the first draft of the resource plan (the POM) and subsequent decisions to give each element a real voice in the process have made PPBS the vehicle these elements use to define themselves and their futures. Indeed, a military department will often speak of its "POM position" as discourse proceeds about alternatives: It is the POM position that defines where that service's leadership has decided to go and how it is going to get there, thereby providing the starting point for the debate of alternatives.

The fact that the secretary of defense can begin the process with a reasonably clean sheet of paper gives him wide latitude to reshape the department as circumstances dictate, albeit at the expense of established programs and priorities. And because the service secretaries run a similar process within their areas of responsibility (as do the heads of the defense agencies, to a lesser extent) they, too, enjoy considerable latitude. From the perspective of the individual program manager, however, this wide latitude can lead to unwelcome turbulence as resources are reshuffled by senior decisionmakers to meet new needs within a relatively fixed budget. Thus, while senior administrators see their ability to shift resources as a strength of the process, operating elements sometimes see this as a serious problem.

Perhaps the area in which this issue arises most sharply and generates the greatest debate is investment, especially the procurement of new articles of equipment. Investment program managers continually complain about the instability and uncertainty PPBS creates for them. A variety of attempts have been made to address this issue, including reviewing stability as an explicit issue in the programming phase, and pilot efforts to manage investment programs through streamlined processes that would (at least in theory) expose them less frequently to review.

One idea DoD has considered would actually promise some programs protection from resource reallocation between development milestones. "Milestone budgeting" would give each program a budget total at each milestone sufficient to carry it to the next milestone (even if several years away), the underlying reason being that these totals can best be estimated at the milestone junctures. Between milestones, these programs would be "off limits" to resource reallocation. From the perspective of the military department secretaries and the defense secretary, milestone budgeting would reduce their flexibility and could lead them to use the operating accounts as a source of funds to meet unanticipated needs. From the perspective of those responsible for the investment accounts, milestone budgeting would promise welcome stability for programs selected (if potentially greater instability for those excluded). It would increase the risk faced by investment managers, however, because it would severely limit their ability to secure added funds between milestones if they found they had underestimated the requirements or if they encountered unexpected technical or other difficulties.

Whatever milestone budgeting's merits, the fact that it is being debated illustrates PPBS's flexible nature and potential ability to shift resources in response to changing circumstances. Because it is a process under the defense secretary's control, it can easily be changed to adapt to new problems or to try new solutions to old problems. This inherent flexibility and adaptability may be why the system has endured for over a generation. Indeed, a careful examination would demonstrate that in each cycle PPBS has been administered, it has been administered somewhat differently than in the prior cycle. Sometimes the differences have been substantial and dramatic (e.g., the introduction of POMs in the late 1960s, and the inclusion of CINC advice in the early 1980s), which is why today's system is very different from the one Secretary McNamara introduced so many years ago.

DECIDING WHAT INVESTMENTS TO MAKE

One of the important clarifications that the 1958 amendments to the National Security Act made in the powers of the secretary of defense related to investment decisions. While these had traditionally been the prerogative of the military services, the amendments confirmed that Congress ultimately held the defense secretary responsible for the department's investment portfolio. Secretary McNamara capitalized on this clarified authority to impose a centralized review of weapons decisions.

Characteristics of that review process included formal documentation of decisions and their rationale, and the use of cost-benefit analyses to weigh the pros and cons of alternative courses of action. Originally resisted by the uniformed leadership, these characteristics are now widely accepted within the defense community—much more so than elsewhere in the federal government.

For major systems—i.e., those exceeding a threshold value for either development or production—the process now begins with a mission needs statement drafted by the responsible party. The system proceeds through a series of milestones, overseen by the Defense Acquisition Board (DAB), which is chaired by the under secretary of defense for acquisition, technology, and logistics with representatives from the military departments and the joint staff. The DAB's approval is required to enter each milestone phase (concept and technology development, system development and demonstration, and production and deployment).

David Packard, deputy secretary of defense from 1969 to 1971, began the practice of gathering advice on milestone investment decisions through an organized board, creating the Defense Systems Acquisition Review Council (the DAB is its contemporary successor). Packard also initiated the concept of independent cost estimates being produced by the secretary's office as a check on what he considered the too often optimistic views of program managers. DoD's reasonably good record in estimating future costs of technologically ambitious systems (which is much better than that of most federal agencies, and better than that of many large-scale private undertakings) owes a great deal to this innovation.

Perhaps the most important milestone decision for DoD is whether to proceed with production. Development expenses are usually a modest fraction (typically 20 percent) of a system's total acquisition cost. Thus, the financial burden of a development decision does not loom nearly as large as that of a production decision. Moreover, in development there is always the hope that further research will resolve any difficulties the system has encountered. The production decision involves an acceptance of the article as worth the department's investment funds. For these reasons, and because of its long-standing distrust of DoD's decisions to proceed with systems, Congress mandates that systems pass an independent test before procurement in quantity begins.[4]

DoD completely separates development from operational testing. Such tests are expensive, however, so the department is typically reluctant to spend the funds necessary to achieve high confidence in the test results.

Before and during World War II, a substantial amount of weapons production took place in government factories—arsenals or shipyards. This is no longer the case. Government-operated shipyards and depots are still responsible for much of the maintenance work on military systems, including major overhauls, but weapons systems are produced by private companies (sometimes using facilities and/or equipment still owned by the government).[5] Thus, one of DoD's important managerial decisions is how—if at all—it wishes to intervene in the marketplace to shape the set of suppliers that bid on its work. Similarly, the contractual relationship between the govern-

---

[4]This requirement was initiated by Congress in the 1970s.

[5]Congress mandates the minimum proportion of maintenance work that must be carried out in government-owned and -operated facilities (currently 50 percent).

ment and the private contractor is a critical DoD administrative decision.

Defense contracting takes place under the Federal Acquisition Regulations (FAR), a regulatory code that governs all procurement by the U.S. government. A key philosophical tenet of this code is full and open competition, a factor that has an important bearing on the incentives faced by contractors, the way in which procurement is carried out, and the government's ability to intervene directly to shape the marketplace. The FAR also embodies a variety of social policy decisions that the federal government insists be reflected in its acquisition practices, most notably support to small business and special consideration for the disadvantaged entrepreneur.

Private contractors (rather than government labs) also typically carry out the development of major systems. The U.S. military acquisition system has evolved such that the firms that undertake development also undertake production. It is difficult to compete in the procurement of a system developed by another contractor,[6] so, in general, the competition for development effectively becomes the competition for production.

Development of a system that pushes the technological frontier is risky, and private firms understandably wish to limit their exposure to such risks when undertaking development contracts. Private firms thus seek cost plus (or similar) contracts for development, while promising in various ways to hold down the costs of production. As a practical matter, this leaves the government bearing not only most, if not all, of the risk, but also the embarrassment if the risks prove greater than the contractor's estimate. In several periods, DoD attempted to limit its risks (total package procurement in the 1960s, fixed price development contracts in the 1980s), but each attempt was abandoned after being perceived as creating problems worse than the ones it was meant to solve.

All this has left DoD in an unsatisfactory situation. Competition at the development stage, when so little is known, encourages contractors to overpromise on performance, especially because they know

---

[6]Difficult but not impossible. In the 1980s, DoD ran several production competitions, called "second sourcing."

that securing the development contract virtually guarantees them the production contract, where most of the profit potential resides. Then, when the government recognizes that the promises were inflated, it usually must face one of two unpalatable choices: delay the program substantially to switch to another provider, or accept a substantial restructuring of the program with the current contractor.

And firms overpromise not just on performance; they often overpromise on schedule, as well. Schedule delays plus the additional time consumed by program restructurings to resolve performance shortfalls can produce substantial delays in fielding relative to initial expectations. From the contractor's point of view, the federal government can be a capricious client, changing the performance specifications in the midst of development and thus necessitating contract renegotiations and program restructuring.

A different difficulty is created because of industry's belief that research and development (R&D) is not profitable:[7] the extent of research not directly funded by the government itself is sharply limited, and thus so is the set of choices available to DoD. Cold War budgets could support a large government-funded R&D program; but even though R&D has been somewhat protected in the post–Cold War drawdown, the budgetary appetite of a few large programs has limited the investment in innovation.

These recurring acquisition difficulties explain the perennial call for acquisition reform. The Clinton administration was no exception; nor is the current Bush administration likely to be so.

The Clinton administration's approach to acquisition reform was led off by Secretary of Defense William Perry's 1994 paper, "Acquisition Reform: A Mandate for Change." In it, Perry emphasized the loss to DoD from its alleged inability to acquire state-of-the-art commercial technology, which, he asserted, reflected the difficulties created by

---

[7]Indeed, it appears that many of the big R&D contracts of the Cold War earned, at best, subnormal profits, and some R&D competitions explicitly stipulated a company "investment" (e.g., the F-22) that was to be repaid through production profits. Thus, one important academic economist characterizes U.S. weapons procurement as a competition for "production prizes," with firms vying to subsidize the R&D phase (see William Rogerson, "Economic Incentives and the Defense Procurement Process," *Journal of Economic Perspectives*, Vol. 8, No. 4, Fall 1994, pp. 65–90).

the FAR and by DoD practices for doing business with commercial companies. The administration helped develop the Federal Acquisition Streamlining Act of 1994, which encouraged the purchase of commercial products whenever possible and eliminated government-unique certification and accounting requirements, especially for smaller purchases. The Act did not, however, change the principle of fair and open competition.

One of the most significant steps taken to create a more commercial environment was the decision to replace military specifications and standards. The traditional procurement process had typically relied on government-unique specifications and standards, but by 1997, several thousand military specifications and standards had been canceled or replaced by performance specifications or, when practicable, nongovernment standards.

Moving beyond these steps, the Clinton administration's second under secretary of defense for acquisition and technology, Paul Kaminski, sought to directly attack the related problems of how long it takes to acquire a system and how much systems cost. Two innovations particularly sought to change the governance of defense acquisition:[8]

- Cost as an independent variable (CAIV). The aim of this innovation was to reduce life-cycle costs by making cost a driver in system design (replacing the Cold War emphasis on performance). The CAIV picks a cost objective and focuses on cost-performance tradeoffs to achieve savings.

- Advanced concept technology demonstrations (ACTDs). The purpose of ACTDs was to shorten the acquisition cycle (and improve performance) by moving directly to fieldable prototypes. Evaluations were to take place in the field and were to be carried out, in part, by the users of the technology, who could explore how new capabilities might be used and recommend adjustments to improve system performance before a full acquisition decision was made.

---

[8]For a full view of Kaminski's efforts, see his prepared statement for the House Committee on National Security, "Defense Acquisition Reform," February 26, 1997, DoD Testimony, 105th Congress, first session, 1997.

These innovations were the latest in a long series of acquisition reforms that began in the Hoover commissions (1949 and 1955) and continued through the Fitzhugh commission (1970), the Commission on Government Procurement (1972), the Grace commission (1983), and the Packard commission (1986).[9] Indeed, it can be argued that defense procurement has steadily improved over the past five decades, as measured by acquisition results (e.g., the performance of equipment in combat operations). But further improvement is still highly desirable, as can be seen in the call of Kaminski's successor, Jacques Gansler, for DoD to concentrate on cutting in half the time it takes to acquire weapons systems.[10] Perhaps acquisition reform is best seen as an evolutionary, rather than revolutionary, set of changes.

DoD has also taken advantage of "other transactions" procurement authority, increasing its flexibility because fewer regulations apply to "other transactions." This form of procurement authority was originally granted to the Defense Advanced Research Projects Agency (DARPA) by the 1994 Defense Authorization Act; it was then extended to all of DoD on a trial basis by the 1997 Act. Major weapons systems that have benefited from this initiative include the Navy's Twenty First Century Destroyer Program (DD21) and the Air Force's Evolved Expendable Launch Vehicle (EELV). The hope is that the flexibility of "other transactions" authority will translate into lower costs, but it is too soon to tell yet.

ALTERNATIVE APPROACHES TO DECISIONMAKING

Ever since McNamara's tenure as defense secretary, DoD has emphasized the principle of optimization in making decisions. Optimization requires a clear statement of objectives against which the benefits and costs of alternative courses of action are weighed. The Cold War was well suited to this decisionmaking paradigm. Not only was the opponent well known, but the threats the opponent posed

---

[9]For a useful summary of these earlier reform recommendations, see Defense Policy Panel and Acquisition Policy Panel, House Committee on Armed Services, *Defense Acquisition: Major's Commission Reports*, Committee Print No. 26, U.S. Government Printing Office, Washington, DC, 1988.

[10]*Defense News*, 6 September 1999, p. 1.

were well specified. Scenarios could be devised that allowed planners to optimize their forces and programs against the threats the Soviet Union presented. Indeed, it can be argued that the U.S. success in winning the Cold War with a defense effort that represented a gradually shrinking economic burden (relative to the output of the U.S. economy) is a tribute to the effectiveness of optimization as a decisionmaking principle: it allowed available resources to be used wisely.

With the end of the Cold War, however, the decisionmaking environment changed significantly. It was no longer clear who would challenge the United States in a way that would require the use of military force, and it was no longer clear what scenarios should be used to judge the effectiveness of future forces. For the greater part of the Clinton administration, DoD focused on having to conduct two nearly simultaneous MTWs—one in Iraq and one in North Korea—as a useful benchmark. But the threat of Iraq or North Korea invading its neighbor began to appear less likely (and their forces less capable) over the course of the 1990s. Regardless of the ill intent of these two countries (and others), however, it was becoming clear that optimizing U.S. forces to fight two nearly simultaneous MTWs was impeding what was really needed: a transformation of U.S. forces to enable them to cope with the wide variety of new, often unpredictable challenges they were facing. Thus, the question became: Given great uncertainty about what opponents the United States might face in the future, and about the location and nature of the conflicts U.S. forces might face, should classic cost-benefit optimization still be the guiding principle for decisionmaking?

Cost-benefit analysis came out of an effort by social scientists to apply the tools of economics to Cold War military problems, and the same tool kit may also yield instruments suited to the uncertain post–Cold War world.

One of the basic principles of economics is that diversification of one's portfolio—i.e., hedging against a variety of possible outcomes—is the appropriate investment strategy when faced with uncertainty. To apply this principle to contemporary defense problems, one must identify the set of possible situations—i.e., the set of possible futures, or even future scenarios—for which the United States should be developing hedges. Since the forces and programs appropriate for one future will not necessarily be appropriate for another, it is only with the greatest good luck that a program optimized for one future will reasonably cover the bets that need to be made. The issue thus becomes how many bets to make and how large they should be, not which is the best single wager.

Warning and the rapidity with which U.S. military forces respond to unforeseen (and currently unforeseeable) circumstances form another approach to an uncertain security environment. If warning can be sufficiently timely or forces are sufficiently flexible in responding to the unexpected, a single programmatic solution to the defense problem might once again be appropriate. Improvement in the ability to discern and act on warning and in the flexibility with which forces respond to unforeseen circumstances can substitute for some of the "portfolio diversification" of the capabilities resident in U.S. forces that would otherwise be required to handle the uncertainties of the post–Cold War world.

ALTERNATIVE DECISIONMAKING STYLES

Decisionmaking gradually became more centralized, consistent with the Cold War emphasis on optimization and with the strengthening of the defense secretary's powers that the 1949 and 1958 amendments to the National Security Act brought. The trend accelerated during Secretary McNamara's tenure, in the 1960s. The secretary of defense's office took control of the defense program, raised issues concerning service plans, and forged a set of decisions that emphasized rationalization of defense efforts in accord with optimizing criteria.

Central planning as a decisionmaking style was well suited to the challenge posed by the Cold War. It also solved the decisionmaking problem of any organization operating in a nonmarket environment in that it created a way to produce coherent and internally consistent decisions when there were otherwise no "signals" to guide the organization's constituent elements. That said, however, central planning can also be stifling. In the uncertain post–Cold War world, it could restrict the very innovation needed to generate alternative options with which to hedge DoD's bets.

As an alternative, the central actors in the department would set the "rules of the game" governing decisions about future forces, their training, their equipment, and their preparations for sustained operations, but would then leave much of the decisionmaking about particular choices to the department's constituent elements. Such decentralization would be a significant shift in the managerial paradigm, but not as significant as might first be perceived. Training decisions remained largely decentralized even in an era of centralized decisionmaking, and in many ways the POM process has already operated in a decentralized manner since the late 1960s. But it would be a significant change for the management of investment decisions.

While a more decentralized decisionmaking approach might encourage innovation, it is not without its own problems. It could restrict the ability of central actors to redirect the department's activities in response to fast-changing circumstances. If authority is decentralized, so is responsibility, which may be inconsistent with Congress's preference to hold the top leaders of the department accountable for the department's actions.

Decentralization would also run up against the question of how to manage the defense agencies. Created over many years (starting in the 1950s), these agencies typically arose because of the perception that it would be more efficient (or at least more effective) to have a consolidated organization carry out certain functions than to permit each military service its own capability. Many of DoD's intelligence, research, and support functions were being carried out by these centralized organizations, and that structure would have to be reconsidered in any serious effort to decentralize.

A very different management issue relates to the time focus of decisionmakers' attention. Especially since Secretary McNamara's tenure, the defense secretary's focus has been on the future, on looking ahead and planning how DoD should best cope with future events. Surprisingly little attention is paid to how well the department executes the plans it has on the books. Such attention as is paid to execution is disproportionately concentrated in the very top of the department and focuses on programs that are "in trouble." The routine monitoring of program execution is largely left to the secretary's assistants, and to the military departments and defense agencies—in essence, a style of management by exception.

This is, of course, a sweeping generalization, and there are exceptions. Secretary Weinberger, in the early 1980s, instituted a series of secretarial performance reviews in which each military department reported directly to him on the performance of major programs once a month. He used these to monitor such issues as the health of the All-Volunteer Force and the president's strategic nuclear modernization program. In the late 1980s, Deputy Secretary William Taft experimented with a biennial POM process, using the off-year to conduct an execution review. Neither innovation survived much beyond its creator's tenure.

It is often observed that the troops on whom the boss checks frequently are the troops who do the best. Especially if a future secretary were to move to a more decentralized decisionmaking style, it would be useful to balance that style with increased attention to program execution. Such attention would have the added benefit of helping DoD more promptly resolve problems that arise—for instance, the concerns over reduced readiness due to reduced resources that existed before the Bush administration's major increases in defense spending. It would also help the department understand how the results of experiments and real-world operations inform decisions about the very uncertain future DoD faces in the years ahead.

LOOKING TO THE FUTURE

Looking to the early years of this century, two of the four enduring decisions any military organization must make are likely to continue being the focus of debate. First, despite the sharp increases in the defense budget that the Bush administration has made, the size and the structure of U.S. military forces are once again likely to be debated. Indeed, this issue has never been fully joined, despite the several major reviews conducted in the 1990s. The current force structure therefore represents more of an evolution than a definite choice geared to the realities of the present era. The strains placed on the current force by contemporary operations, plus the new missions in counterterrorism and homeland security, only add to the pressure to consider this issue explicitly.

The second enduring focus of debate is the nature of equipment that should be used to arm the U.S. force. The urgent need to recapitalize the present generation of equipment has put this decision high on the department's agenda. In short, the new century brings to the fore a critical set of specific defense decisions that must be made, and with them comes the opportunity to consider afresh *how* those decisions will be made. It will be a fascinating period for both students and practitioners of defense decisionmaking.

Chapter Two

RESPONDING TO ASYMMETRIC THREATS
Bruce W. Bennett

As Chapter One indicates, Cold War planning dealt largely with "symmetric" threats—strength-on-strength planning vis-à-vis the Warsaw Pact, especially in central Europe. Warfighters find it easiest to address such threats, ones they understand thanks to the Cold War experience. But America's ability to prevail handily against symmetric threats has forced U.S. adversaries to pursue asymmetric threats. In one sense, strategies often include asymmetric components in that they seek to exploit the other side's vulnerabilities, but the Cold War's image of two broadly similar superpowers obscured that fact.

The 1997 Quadrennial Defense Review (QDR) identified asymmetric threats, or "challenges," as a major issue for the U.S. military.[1] Previous RAND work[2] defined asymmetric threats as those that attack vulnerabilities not appreciated by the target or that capitalize on the target's limited preparation against the threat. These threats usually rely on concepts of operation (CONOPs) that differ from the target's and/or from those of recent history.[3] The U.S. military understands

---

[1]William S. Cohen, *Report of the Quadrennial Defense Review*, Department of Defense, May 1997, in particular pp. 4 and 49–51, but also pp. vii, 12, 13, 19, 41, and 43.

[2]Bruce W. Bennett, Christopher P. Twomey, and Gregory F. Treverton, *What Are Asymmetric Strategies?*, DB-246-OSD, RAND, 1999.

[3]Additionally, asymmetric threats can serve political or strategic objectives not shared by the victim. For example, in 1941, U.S. economic sanctions against Japan were intended to coerce Japan into stopping its aggression in East Asia. But the Japanese, having different strategic objectives, responded with an asymmetric strategy: a strike against Pearl Harbor that sought to neutralize the U.S. Pacific Fleet and thereby to convince the United States to disengage from East Asia.

its own strengths and tends to focus on them even when it does not see comparable preparations by a prospective adversary, assuming that warfare will be largely symmetric in character. This failure to adequately recognize asymmetric threats is neither new nor unique to the U.S. military.[4] However, given that asymmetric threats are now the greatest threats to the U.S. military and U.S. society (as the terrorist attacks of September 11, 2001, drove home), the subject requires particular attention in U.S. military planning.

This chapter characterizes asymmetric threats and outlines steps the U.S. military needs to take to counter them. It starts with a general introduction to asymmetric threats, explaining why and how they would be wielded. It also addresses the importance of adversary surprise and anonymity, as well as the nature of challenge and response cycles. It then focuses on threats based on chemical and biological weapons, which are a significant element of the current threats to the United States and its allies.

FROM THE COLD WAR TO THE PRESENT

The two principal arenas for Cold War military confrontation with the Soviet Union—the NATO Central Front on the inter-German/Czech border, and strategic nuclear conflict—were perceived by the United States as symmetric confrontations, creating a culture of expectation for strength-on-strength combat. Yet even these Soviet threats included many elements that were asymmetric. On the Central Front, armor faced armor in a contest of maneuver, and both sides depended on artillery, aircraft, and naval forces. U.S./NATO strategy eventually focused on developing technologically superior weapons to provide an asymmetric advantage for NATO. Meanwhile, the Soviets relied on quantity and strong CONOPs, including the heavy use of special forces and chemical and biological weapons (CBW); artillery disruption leading to breakthroughs, especially against the weaker forces of some NATO allies; and penetration concepts, such as operational maneuver groups.[5]

---

[4]Comparable historical concepts have included the writings of Sun Tzu, maneuver warfare, and centers of gravity.

[5]An operational maneuver group was a corps-sized heavy maneuver force designed to penetrate an operational breakthrough (the collapse of at least part of a NATO corps

Despite the efforts of some military specialists, however, the extent of asymmetry in the Soviet threat was not well comprehended in the United States.

Each side's strategic nuclear forces consisted of a triad: intercontinental ballistic missiles (ICBMs), submarine-launched ballistic missiles (SLBMs), and bombers. Many in the United States argued that stability was maintained by the assured ability to destroy opposing urban/industrial areas even after suffering a first strike against strategic weapons. The United States planned a nuclear response to a major Soviet conventional force breakthrough in NATO, and proposed a doctrine of limited nuclear operations to control the escalation from such nuclear weapons use. After the Cold War, the United States learned that the Soviets had taken asymmetric approaches to many of these U.S. concepts and did not accept the primacy of assured destruction. For example, the Soviets created vast production capacities for and stockpiles of such biological weapons (BW) as anthrax and smallpox and planned to use them against the United States.[6]

This perception of symmetry led many U.S. analysts to conclude that quantitative measures of military hardware, units, and warfighters were the key metrics for evaluating the military capabilities of the United States and the Soviet Union. Analysts compared the numbers of U.S./NATO tanks, artillery pieces, combat divisions, military manpower, and fighter aircraft to Soviet/Warsaw Pact equivalents.[7] Strategic analysts counted ICBMs, SLBMs, bombers, and warheads, and assessed mixed quantitative/qualitative measures such as equivalent megatons and countermilitary potential.[8] Analysts and de-

---

sector) and overrun targets behind the front, including headquarters, airfields, major storage depots, and even political targets. The hope was that by so doing, these groups would expedite the strategic collapse of NATO's defenses.

[6]See Kenneth Alibek, "Biological Weapons," presented to the USAF Air War College, November 1, 1999. It refers to Soviet storage of anthrax in excess of 100 tons, with an annual production capacity of thousands of tons; it also says that the Soviets had stockpiles of plague and smallpox that were each roughly 20 tons.

[7]See, for example, *NATO and the Warsaw Pact: Force Comparisons*, NATO Information Service, 1984; and William P. Mako, *U.S. Ground Forces and the Defense of Central Europe*, The Brookings Institute, 1983.

[8]See, for example, Paul Nitze, "Considerations Bearing on the Merits of the SALT II Agreements as Signed at Vienna, *The Congressional Record—Senate*, July 20, 1979, pp.

cisionmakers talked in terms of a "military balance," as if roughly equal force quantities were stable and greater quantities conferred military advantage.[9] In practice, most comparisons concluded that because the Soviet Union had more military equipment, personnel, and units, it held a military advantage, though there were some uncertainties even about these quantities. These comparisons were largely strength-on-strength comparisons and thus tended to ignore other key characteristics and relative vulnerabilities—the essence of asymmetric threats.[10] In addition, these analyses seldom considered whether U.S. and allied forces could perform their missions or meet their operational requirements.

Today, military leaders and analysts, trained during the Cold War, are tempted to apply similar methods in discussing U.S. conventional and nuclear superiority over potential adversary states.[11] Figure 2.1 shows how U.S. conventional and nuclear capabilities overwhelm the conventional capability of most prospective adversaries,[12] especially the "rogue states," such as Iraq, North Korea, and Libya. And the addition of weapons of mass destruction (WMD) by these rogue states is not seen as providing sufficient "strength" to offset U.S. ca-

---

S10070–10082; and Robert L. Leggett, "Two Legs Do Not a Centipede Make," *Armed Forces Journal International*, February 1975, pp. 30–32. A critique of these methods is contained in Bruce W. Bennett, *Assessing the Capabilities of Strategic Nuclear Forces: The Limits of Current Methods*, N-1441-NA, RAND, June 1980.

[9]See, for example, arguments on the Conventional Forces in Europe (CFE) negotiations in James R. Thomson and Nanette C. Gantz, *Conventional Arms Control Revisited: Objectives in the New Phase*, N-2697-AF, RAND, December 1987.

[10]These comparisons acted as if the relative quality of military personnel or command-and-control or intelligence or even (in many cases) hardware quality did not matter in the overall assessment. The Soviets included qualitative force multipliers in their assessments, as discussed in Allan S. Rehm and Joan F. Sloan, *Operational Level Norms*, SAI-84-041-FSRC, SAIC, April 24, 1984, especially Section 3.

[11]Most military analysis is still fixed on repelling ground force invasions. Neither the diversity of other U.S. military engagements nor the differences in each side's vulnerabilities and force requirements get much attention. Although such threats as ballistic missiles are recognized, conflicts that entail their use per se (rather than as ancillaries to an invasion) are downplayed.

[12]The United States explicitly wishes to avoid becoming an adversary of the world's other great powers. "The United States is committed to expanding its network of friendships and alliances with the aim that eventually all of the world's great powers will willingly cooperate with it to safeguard freedom and preserve peace" (Donald H. Rumsfeld, *Guidance and Terms of Reference for the 2001 Quadrennial Defense Review*, June 22, 2001, p. 1).

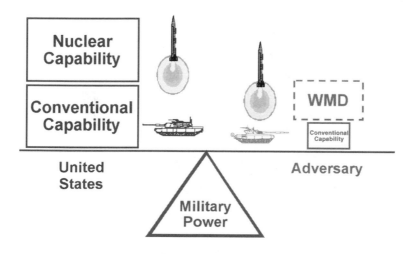

Figure 2.1—Symmetric View of Military Power

pabilities. During the Cold War, many smaller states accepted this imbalance because the Soviet Union counterbalanced U.S. strength. But now they face the United States on their own and cannot afford the U.S. level of military capabilities.

U.S. conventional military superiority remains focused on its traditional threats: armored and air assaults over relatively open terrain to conquer territory of a U.S. ally. In the 1990s, theorists spoke of a revolution in military affairs (RMA) that would allow the United States to use technology to overwhelm such a threat. "An RMA . . . renders obsolete or irrelevant one or more core competencies of a dominant player."[13] Indeed, the United States has made most opposing forces of armor, aircraft, and ships obsolete in symmetric conflicts against it.

Figure 2.2 suggests that because adversaries cannot achieve a military balance with the United States using symmetric approaches, they have been induced to find other ways to undermine U.S. mili-

---

[13]Richard O. Hundley, *Past Revolutions, Future Transformations*, MR-1029-DARPA, RAND, 1999, p. 9.

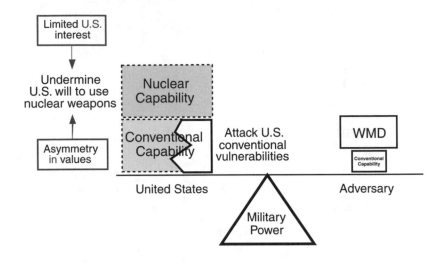

Figure 2.2—How Adversaries Might Use Asymmetric Threats

tary power—by attacking U.S. military vulnerabilities or America's will to use its military might.[14] Any of the approaches could remove a significant part of U.S. strength from the balance, giving adversaries a chance to prevail.

Few U.S. adversaries now contemplate a conventional force invasion of a neighboring territory, especially one allied with the United States, but they have still sought to improve their military capabilities.[15] They are also pursuing asymmetric approaches to achieve their objectives, such as standoff coercion, guerilla warfare, subversion, and information warfare. The little "RMAs" that play a role in the pursuit of these approaches include cover, concealment, and deception to prevent the United States from recognizing threats or

---

[14]This very simple depiction ignores the role of U.S. allies and U.S. forward basing. Figure 2.3 extends this simple approach.

[15]One exception, North Korea, may yet invade its neighbor, South Korea, using conventional forces. Its asymmetric "facilitating force," which would remove or suppress many U.S. and South Korean defenses, includes artillery with chemical shells that would be used against U.S. and South Korean ground forces; special forces and ballistic missiles armed with CBW to suppress airfields, ports, command, control, communications, and logistics; and cruise missiles and aircraft to sustain CBW contamination.

countering them; new delivery means, such as ballistic and cruise missiles or unmanned aerial vehicles (UAVs) that are inexpensive and difficult for the United States to counter; and WMD to affect military operations or coerce neighbors.

EXAMPLES OF ASYMMETRIC THREATS

Examples of asymmetric threats might include the following:

- Computer hackers use e-mail viruses to destroy U.S. military personnel records and the software used to process them, thereby seeking to delay U.S. force deployments and mobilization.

- Terrorists explode bombs against civilian targets in New York City.[16]

- Adversary special forces fire handheld surface-to-air missiles (e.g., SA-16s) against U.S. cargo aircraft, tankers, and command-control aircraft taking off from theater airfields.

- Operating from fishing ships, Iraqi special forces spray BW upwind of U.S. Navy ships in the port of Jabal Ali in the United Arab Emirates. (Jabal Ali is the largest port in the Persian Gulf.)

- Seeking to split the U.S.-led coalition against Iraq, Saddam Hussein claims that U.S.-sponsored sanctions are starving Muslims in Iraq.

- North Korea uses chemical weapons (CW) against the Republic of Korea.

- China threatens a nuclear attack on U.S. cities if the United States interferes in its actions against Taiwan.

These threats may employ novel weapons but need not do so; the weapons can be similar to those of the target. What makes threats asymmetric is the difference in CONOPs and that such threats are used against the target's unexpected vulnerabilities. The target is usually surprised, and the stun effect may delay a response—all of which amplifies the impact. It takes years to build appropriate mili-

---

[16]This example was in the first draft of this text, which was prepared well before September 11, 2001.

tary forces and capabilities to counter such threats. Asymmetric threats are relative; some are more asymmetric than others.

HOW WOULD ADVERSARIES SHAPE ASYMMETRIC THREATS?

Figure 2.3 refines the concepts of Figure 2.2. September 11 notwithstanding, most of the possible conflicts in which the United States will want to intervene will be far from U.S. shores. Hence, the United States must project military power into a region, which requires military means, U.S. will, and the will and support of regional allies that will allow U.S. force and logistics basing and often fight alongside the Americans. U.S. discussions of asymmetric threats usually focus on adversary operations against U.S. military capabilities, such as Scud missiles with chemical warheads being fired at airfields to degrade aircraft operations or disrupt the flow of combat aircraft and their support into the theater. Airfields and other combat support facilities make good targets because they are far more vulnerable than combat aircraft themselves. The United States tends to concentrate its military resources in a few locations in the theater (e.g., airfields, ports, and command facilities) to minimize costs and coordination efforts, but such "massed" facilities make excellent targets for WMD, as well as for conventional weapons delivered by special forces.[17] Despite the recent U.S. focus on protecting these facilities, they remain vulnerable to at least some asymmetric threats.

However, adversaries may find it easier and safer to undermine the will and support of key U.S. regional allies than to attack U.S. military capabilities. Allies understand that their military forces and civilians tend to be more vulnerable than U.S. deployed forces. Their interests frequently differ from those of the United States, especially if they are not the immediate targets of an adversary's invasion. In 1990, when Iraq's invasion victimized only Kuwait, other countries in the Persian

---

[17]An adversary's military culture may lean toward asymmetric threats. For example, because most of North Korea's political and military leadership came out of the special forces Kim Il Sung operated in World War II, North Korea has emphasized special forces operations. By contrast, most U.S. military analysis expresses concern over special forces operations but largely ignores the damage they can do and thus devalues potential actions to counter them.

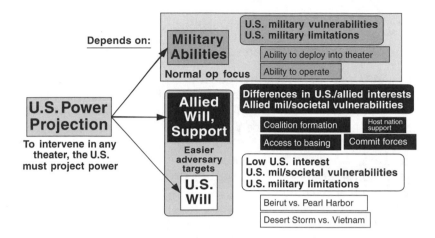

Figure 2.3—U.S. Requirements for Military Power Projection

Gulf were initially uncertain about how much access and support to give the United States. Each worried about its vulnerability to Iraq, fearing that allowing U.S. forces to attack Iraq from its soil might lead to Iraqi retaliation. That is, in fact, what happened: Iraq responded with Scud missiles.

Even when the United States can gain access to regional states, to operate it needs "host nation support"—workers for docks and air-fields; power, water, communications, and transportation infrastructure; and so on. Unless these workers are protected against WMD threats by either the host nation or the United States, they may become casualties or flee. Either would degrade U.S. deployment and operations.

U.S. will is especially at risk because U.S. interest in recent and prospective regional conflicts far from home has been limited. U.S. military casualties caused a collapse of U.S. will in the Beirut, Lebanon (1983) and Mogadishu, Somalia (1993) interventions. Some foes contemplate how the U.S. will to intervene, especially in offensive roles, such as in Kosovo, can best be defeated—by posing a strategic threat to the United States[18] or by simply raising the cost of U.S. involvement so that it exceeds the advantages to be gained in interventions of low U.S. interest.

Yet calculating the effects that asymmetric threats or attacks will have on U.S. will is no easy task for would-be adversaries. The 1983 attack on the Marine barracks in Beirut led to a U.S. withdrawal from Lebanon. U.S. interest was very low, and the imposed cost—several hundred soldiers killed—was moderate, with the result that the United States decided to discontinue the intervention. However, the 1941 Japanese attack on Pearl Harbor drove the United States into World War II. The United States perceived its interest in Hawaii to be so vital that Japan was unable to impose too high a price for U.S. intervention. Instead, America felt compelled to remove Japan as a threat to U.S. national security. Getting the threat just right requires imposing costs large enough to prevent U.S. intervention but not so large as to trigger a U.S. change in objectives and/or a demand for revenge.[19] After September 11, there was continued concern about losing too many American lives, but there was no doubt that the United States would do what was required to overturn Afghanistan's Taliban regime and attack Al Qaeda.

The statement of a North Korean sympathizer evokes the dilemma would-be adversaries face: "Which is better prepared for nuclear exchange, North Korea or the USA?. . . The DPRK can be aptly described as an underground fortress. . . . For their part, the North Koreans are

---

[18]See, for example, Northeast Asia Peace and Security Network, "DPRK Report #19: The Importance of NK Missiles," August 9, 1999, http://www.nautilus.org/pub/ftp/napsnet/russiadprk/dprk_report_19.txt. In addition, in an article in the *New York Times*, Michael R. Gordon quotes Chinese official Sha Zukang as saying, "Once the United States believes it has both a strong spear and a strong shield, it could lead them to believe that nobody can harm the United States and they can harm anyone they like anywhere in the world. There could be many more bombings like what happened in Kosovo." Gordon then goes on to say, "He [Sha Zukang] made plain that China's fear was not that the United States would launch a surprise attack on China, but that a missile shield would lead American politicians to believe that the United States was so powerful and well protected that it could act with virtual impunity." (Michael R. Gordon, "China Looks to Foil U.S. Missile Defense System," *New York Times*, April 29, 2001, p. 6.)

[19]Because the United States has learned that many adversary's threats are idle, an adversary may be tempted to demonstrate its capabilities to make its threats credible. Such a demonstration, especially if carried out within the United States, would almost certainly be escalatory.

highly motivated candidate martyrs well prepared to run the risk of having the whole country exploding in nuclear attacks from the USA by annihilating a target population center."[20] Arguably, even the current U.S. interest in Korea would not justify risking a single nuclear attack on the U.S. homeland. But the very first North Korean nuclear attack on U.S. soil would challenge U.S. survival, likely making the United States prepared to withstand great losses to eliminate the North Korean threat.

It is also tempting to think of adversaries posing only a single asymmetric threat—i.e., just one of the examples cited here—even though adversaries may plan some diversity and combination of threats. Then, because U.S. military power requires a combination of many factors, damage to any one of those factors could degrade U.S. combat capability. According to Mark Mateski, the Provisional Irish Republican Army (IRA) once warned Margaret Thatcher, "We only have to be lucky once—you will have to be lucky always."[21] Osama bin Laden might have said the same thing.

THE IMPORTANCE OF SURPRISE AND ANONYMITY

Asymmetric threats target unappreciated vulnerabilities, and they tend to result in surprise. Attacking at unexpected times or in unexpected ways heightens the surprise; so does hiding the preparations or misleading the victim about one's objectives, strategy, capabilities, and deployments. Adversaries that avoid having attacks attributed to them may achieve many objectives *and* avoid retaliation. According to CIA Director George Tenet, "More than ever we risk substantial surprise. This is not for a lack of effort on the part of the

---

[20]Kim Myong Chol, "The Future of the Agreed Framework," Northeast Asia Peace and Security Network, Policy Forum Outline (#23C), November 24, 1998, http://www.nautilus.org/pub/ftp/napsnet/special%5Freports/pf23c%5Fkim%fFon%5Fagreed%5F framework.txt.

[21]Mark Mateski, "The Policy Game," *Red Team: The Journal of Military Innovation*, August 1998, http://www.redteamjournal.com/issuePapers/issue_paper2.htm. In the same article, Mateski says, "And while we spend our strength chasing Bin Laden, each potential adversary will continue to prepare to fight us on its own terms, whatever terms suit it best. For Iraq, the preferred way of war may be another ground assault supplemented with chemical and biological weapons. For North Korea, it may be a massive frontal onslaught. For China, it may be a cat-and-mouse maritime contest. In any case, we must be prepared for them all."

Intelligence Community; it results from significant effort on the part of proliferators."[22]

The possibility of surprise is increased by the fact that U.S. planning and analysis remain so dominated by high-end, largely symmetric threats that resemble those of the Cold War. Most defense policy-makers and analysts lack a strategic appreciation for asymmetric threats (such as CBW threats in Korea before 1997), as well as a tactical/operational appreciation for them. They may have known, for instance, that terrorism was a threat in Saudi Arabia before the bombing of Khobar Towers in 1996, but not when and how the towers would be struck. Why the failure?

- U.S. threat-based planning is very susceptible to adversary deception. As a result, the United States mischaracterizes and often underestimates the adversary's threat and thus feels increased surprise when the threat is carried out.

- Analysts tend to "mirror image" an adversary when intelligence on the adversary's strategy and CONOPs is thin.

- Large bureaucracies, plagued by groupthink, have trouble accomplishing the "thinking outside the box" needed to understand how an adversary would employ a threat in novel ways. This also stifles projections of the effects an asymmetric threat may have and thus limits options for changing operations and forces to respond to them.

- Resource constraints tend to focus on force modernization in terms of traditional weapons, such as fighter aircraft, destroyers, and artillery. Less attention is paid to developing and fielding the

---

[22]"Text: DIA Director Tenet Outlines Threats to National Security," U.S. Department of State, International Information Programs Internet Site, Washington File, 21 March 2000. Tenet then cited four reasons why: "First and most important, proliferators are showing greater proficiency in the use of denial and deception. Second, the growing availability of dual-use technologies is making it easier for proliferators to obtain the materials they need. Third, the potential for surprise is exacerbated by the growing capacity of countries seeking WMD to import talent that can help them make dramatic leaps on things like new chemical and biological agents and delivery systems. . . . Finally, the accelerating pace of technological progress makes information and technology easier to obtain and in more advanced forms than when the weapons were initially developed."

equipment and forces required to respond to "unproven" asymmetric threats.

Without adequate preparation, the United States will lack the people and equipment it needs to counter new threats, and responding to such threats takes time.[23]

Most adversaries would like to defeat the United States without having to pay for it. They may attempt to avoid reprisal by hiding their strategy and CONOPs and, perhaps, by carrying out attacks covertly. They may not need to claim credit for the damage done, depending on their objectives. If the United States cannot attribute an attack to an adversary, it may lack the will to retaliate. Thus, deterrence will not work well against an adversary that thinks it can avoid attribution.

CHALLENGE AND RESPONSE CYCLES

As the United States and potential adversaries pursue new military capabilities, "challenge and response cycles" result.[24] Thus, as adversaries' threats create new challenges for the United States, the issue becomes one of how quickly the United States can respond. For example, in the Cold War the Soviet Union used the threat of a massive armored invasion to challenge the United States and its NATO allies. Their response was to seek higher-technology armored forces and, especially, air forces that could interdict Soviet armor. The quality of the U.S. response was inadequate until the late 1980s. As it turned out, this anti-armor response developed for the Soviet threat defeated Iraq's 1990 challenge in the Persian Gulf.

This U.S. anti-armor response challenged other potential adversaries, however. Many responded by developing CBW, thus posing

---

[23]Adversaries often fail to understand asymmetric operations and thus fail to exploit them. See, for example, the discussion of Germany's treatment of chemical weapons in World War I in Kenneth F. McKenzie, Jr., "An Ecstasy of Fumbling: Doctrine and Innovation," *Joint Forces Quarterly*, Winter 1995–96, pp. 62–68. Such failures may reduce the overall effect of the operations, but considerable damage may still be done (e.g., Germany's CW attacks in World War I).

[24]Sam Gardiner and Dan Fox originally described this cycle in unpublished RAND work on RMAs.

a new, asymmetric challenge to the United States, which has re-nounced the use of CBW. The United States counterresponded by threatening reprisals, which succeeded in deterring Iraq in 1991, and it has also worked on better defenses against CBW. But these re-sponses have lagged adversaries' CBW challenges—even today, the United States remains relatively vulnerable to CBW attacks. And CBW is only one among a diverse set of asymmetric threats.

Worse, in today's relatively short wars (often only weeks or months long, as opposed to years), each side largely brings to the conflict the forces and capabilities it has prepared beforehand. If U.S. adversaries fight differently and the United States lags too far in the cycle, it may be leaving itself particularly vulnerable.

Its focus on threat-based planning has caused the United States to lag in many challenge and response cycles. Planning that is threat based requires an established threat. When adversaries hide the de-tails of their threats, it can take years or even decades (if ever) to un-cover the details, which puts the United States behind its adversaries.

The capabilities developed as part of the U.S. RMA also introduce new military vulnerabilities. For example, U.S. use of a global posi-tioning system (GPS) allows precise weapons delivery—unless an ad-versary finds ways to jam the GPS in the area. In addition, many U.S. weapons that attack ground forces by scattering warheads over a broad area—a single CBU-97, which contains 40 sensor-fuzed weapons and covers an area of 15 acres,[25] or a SADARM (sense and destroy armor munition), which covers 20 acres[26]—may work well in deserts or behind adversary lines, but they can cause great collateral damage to friendly military vehicles or civilian vehicles intermixed with adversary forces (e.g., at the battlefield front, along adversary penetrations, or in urban areas). If adversaries focus on creating intermixed environments (a "nonlinear battlefield"), they can make it difficult for the United States to use its "superior" weaponry.

---

[25]Glenn W. Goodman, Jr., "Nowhere to Hide," *Armed Forces Journal International,* October 1997, p. 59.

[26]Goodman, "Nowhere to Hide," p. 61.

THE CHALLENGE OF WEAPONS OF MASS DESTRUCTION

A particularly fearsome class of asymmetric strategies involves WMD—nuclear, radiological, biological, and chemical weapons. These weapons can hurt military forces and civilians in great numbers and are thus a significant element of adversaries' threats. Not surprisingly, President Bush focused on WMD in his 2002 State of the Union Address: "Our nation will continue to be steadfast and patient and persistent in the pursuit of two great objectives. First, we will shut down terrorist camps, disrupt terrorist plans, and bring terrorists to justice. And, second, we must prevent the terrorists and regimes that seek chemical, biological or nuclear weapons from threatening the United States and the world."[27]

President Bush specifically identified North Korea, Iran, and Iraq as states with WMD and as being part of an "axis of evil." A CIA report immediately after the president's speech also identified Russia and China as key suppliers of WMD technology, and Libya, Syria, Sudan, India, Pakistan, and Egypt as states acquiring technology relating to WMD and advanced conventional munitions.[28] It has been widely reported that North Korea possesses 2,500 to 5,000 tons of CW, that Iran possesses thousands of tons, and that Russia has had some 40,000 tons of CW and potentially thousands of tons of biological agents.[29]

Although often discussed as if they were a single category of weapons, the various WMD differ greatly, as Table 2.1 indicates. Many of their effects are a function of the wind blowing fallout or CBW aerosols. Some BW are incapacitating but not lethal; others, such as anthrax (which was spread by letters in the wake of September 11), are quite lethal. Many BW, especially those dispersed as sprays, are neutralized in minutes or hours by sunlight and rain; others (e.g., smallpox) are contagious and spread from person to person. Some CW (e.g., VX) persist in liquid form on the ground and become a protracted vapor and contact hazard until they evaporate or are ab-

[27]See http://www.whitehouse.gov/news/releases/2002/01/20020129-11.hmtl.

[28]See http://www.cia.gov/cia/publications/bian/bian_jan_2002.htm.

[29]See, for example, Office of the Secretary of Defense, *Proliferation: Threat and Response*, January 2001, pp. 56–57.

Table 2.1

Comparing Weapons of Mass Destruction[a]

Weapon Type	Size	Lethal Effect	Time to Effect	Area Covered	Potential Fatalities
Nuclear	100 Kt	Blast	Seconds	35 km^2	100,000–320,000
		Fallout	Hours to weeks	~800 km^2	100,000s
Biological (anthrax)	100 kg	Disease	Days	45–300 km^2	100,000–1,000,000+
Chemical (sarin)	1,000 kg	Nerve damage	Minutes	0.7–8 km^2	3,000–80,000

[a]See Office of Technology Assessment, *Proliferation of Weapons of Mass Destruction: Assessing the Risks,* August 1993, pp. 53–54. This reference assumes that nuclear blast effects in excess of 5 psi overpressure are, on average, lethal. The anthrax and sarin areas were offered for three different weather conditions. Based on other sources, these were apparently the areas of 50 percent lethality; some lethality would occur at far greater distances, especially for anthrax. The fallout area was estimated from Samuel Glasstone and Phillip J. Dolan, *The Effects of Nuclear Weapons,* 1977, pp. 427–430. From their table on p. 430, the downwind distance and maximum width were estimated for a 100 rads/hr dose with a 100-Kt weapon; these were then multiplied by each other and by 0.7 to reflect the actual character of the pattern. Potential fatalities are calculated assuming no treatment and 3,000 to 10,000 people living in each kilometer affected, though the fallout and BW effects will likely go well beyond the range of a city and thus much of the area covered will not have a high population density.

sorbed into the ground (after which they can come back out of the ground over a long period of time). Their persistence varies with temperature, wind conditions, and surface absorption rates.

Because of the large areas they affect, these weapons can be force multipliers. To reduce WMD effects, defenders must employ various protections (reasonable ones exist for CBW, but nuclear weapons are harder to defend against). All of these protections reduce the defending force's effectiveness, but adversaries that deploy CBW risk contaminating areas where their own forces are positioned or need to go, which could cause casualties among their own personnel or degrade performance for their forces.

U.S. threats of retaliation against WMD use, such as that made against Iraq in the Persian Gulf War, are often enough to deter adversaries—at least if those adversaries are not more worried about regime survival than about the U.S. threat. However, the large CW inventories in several countries suggest that some adversaries see using CBW as an operational necessity.[30] They would likely use CBW early, to maximize surprise and devastation, rather than using WMD only to avert final defeat. BW are best employed before an invasion begins because of their long incubation periods; early use (e.g., at D–2) would sicken defenders around the time of D-Day. BW could also be used to cause disruption in a crisis, the adversary seeking to wreak damage while avoiding attribution and thereby escaping retaliation. Yet most adversaries are risk averse, so even a small potential of attributing WMD use to them, associated with the expectation of serious U.S. retaliation, should be sufficient to deter BW use in most circumstances.

A FRAMEWORK FOR RESPONDING TO ASYMMETRIC THREATS

Asymmetric threats pose a quandary for the United States because they threaten U.S. and allied forces, civilians, and interests in diverse ways that are difficult and expensive to counter. This section proposes a framework for responding to asymmetric threats, focusing on responses to CBW threats as an example of the more general problem. The framework is based on two response approaches identified in the 1997 QDR: institutionalizing and internationalizing.[31]

To be effective, responses must be institutionalized. This would require the military to recognize and address all forms of policy—doctrine, strategy, and CONOPs preparation; force structure and equip-

---

[30]For example, North Korea is reported to have 2,500 to 5,000 tons of CW (South Korean Ministry of National Defense, *Defense White Paper, 2000*, p. 86), and Iran apparently has several thousand tons of CW (CIA, *Unclassified Report to Congress on the Acquisition of Technology Relating to Weapons of Mass Destruction and Advanced Conventional Munitions*, January 1 through June 30, 2000). These inventories are far greater than would be needed for strategic deterrence and regime survival. Because the U.S./South Korean plan for war calls for defeat and destruction of the North Korean regime as well as military conquest of the country, any war would be a total war from the North Korean perspective. As such, North Korea would have little incentive to withhold weapons early in a campaign. See "KBS Reports Plan to Topple Kim Il Sung," *Washington Times*, March 25, 1994, p. 16; and Ranan R. Lurie, "In a Confrontation, 'North Korea Will Definitely Be Annihilated,'" *Los Angeles Times* (Washington ed.), March 24, 1994, p. 11.

[31]William S. Cohen, *Report of the Quadrennial Defense Review*, 1997, p. 49.

ment development; and personnel management and training—something DoD does not do well today. There is no single solution to CBW threats, no "silver bullet" that defeats even any individual threat component, let alone the totality.[32] Responses to even the related CW and BW threats often differ greatly, complicating DoD's ability to institutionalize responses to such threats. Instead, an integrated defense effort must include two elements of response institutionalization: institutionalization through protection and institutionalization through threat management. It must also include internationalization, which entails extending the two elements of institutionalization so as to coordinate with U.S. allies and coalition partners. These three parts of an integrated approach are described in the following sections.

Institutionalization Through Protection

Protection against CBW effects has four components: attack operations (destroying delivery systems before they can be launched), active defenses (destroying delivery systems and their payloads en route), avoidance (maneuvering around contamination or working from places not contaminated), and passive defenses (protecting from contamination). As shown in Figure 2.4 and discussed next, these four components of response currently have limited effectiveness for defeating CBW.

Most adversary CBW and delivery vehicle stocks are so large that attack operations, especially in the face of adversary cover, concealment, and deception efforts, would take a long time to make a large dent in them. Active defenses work well against some CBW delivery means, such as combat aircraft and naval ships. Yet very short-range ballistic missiles (e.g., the CSS-8) can fly under current defenses, and longer-range ballistic missiles (e.g., the NoDong) descend at angles and speeds that pose problems for current systems.

---

[32]That an anthrax vaccine suffices for that specific threat may be largely true (even if some share of the victims still die). But this "solution" ignores the challenge and response cycle: eventually, someone may well defeat the anthrax vaccine. Therefore, the United States needs to seek redundant, robust defenses.

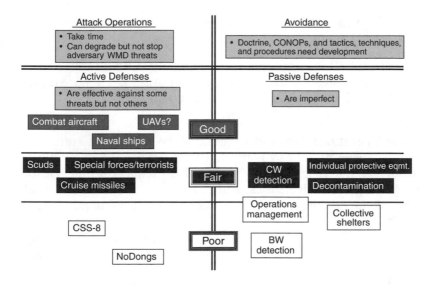

**Figure 2.4—Shortcomings of Today's Means of Force Protection
Against CBW**

Although CBW contamination avoidance can help, the related doctrine, CONOPs, and procedures need more work. Passive defenses against CBW threats provide a fair degree of protection, but only in some cases. Note that attack operations and active defenses work together with avoidance and passive defenses by reducing the amount of chemical and biological material that arrives in the area being defended. The United States is improving its capabilities in all these areas.[33]

All military personnel need to be trained in how to operate in CBW threat and contaminated environments. Today's training is hampered by inadequate doctrine, CONOPs, and procedures; it deals with very short-term chemical threats and tends to neglect biological threats and longer-term chemical threats (partly be denying the latter will take place). Training needs to be broadened to cover the range of CBW threats and to become more universal (all personnel should be

---

[33]See, for example, the description of the new U.S. protective suit technology in Curt Biberdorf, "CB Protective Field Duty Uniform," *CB Quarterly*, March 2001, pp. 17–18.

trained) so that military personnel think in terms of fighting in a CBW environment.

Similar limitations apply in the case of other asymmetric threats. Protecting against the use of nuclear weapons is even more difficult; it is also difficult to protect against terrorism or attacks on information systems.

Attack Operations. One way to reduce CBW threats is to attack their delivery systems, although large CBW stockpiles and large numbers of delivery systems mean that eliminating them takes time. Destroying CBW on the adversary's territory at least makes the foe suffer most of the consequences, but successful attacks require capable munitions and good information about the threats: what and where they are, and how best to destroy them while controlling collateral damage. Attacking CBW preemptively is difficult because of the risk of collateral damage that can occur in areas surrounding CBW facilities. U.S. planning normally anticipates that attack operations against adversary CBW will not start until the adversary has used some CBW, but this approach requires that the CBW attack be correctly attributed. The United States needs to show that the adversary is, in the end, responsible for the damage because it used CBW first.

For example, say the United States plans to use radar to follow ballistic missile tracks back to their launchers, the goal being to kill launchers before they can pack up and move on. Since most missile launchers have in storage five to 20 missiles, destroying launchers means potentially preventing the launch of multiple missiles. If each launcher in a 50-launcher force has a 20 percent risk of being killed each time it launches (a high level of effectiveness), after 10 launches each (over five to 10 days?), the force will be cut down to five (which is enough to still do some damage). Thus, attack operations take days or weeks to have much effect; they do little to attrite the threat early unless there are only a few launchers and they are left out in the open. Still, anything that reduces launch rates relieves pressure on active defenses (and limits the odds of their being saturated).

Active Defenses. The next component of response against CBW threats is interception of CBW delivery systems en route. Active defenses can include ballistic missile defenses, air defenses, naval defenses, border guards, and custom agents. They must be matched to the threatening delivery system(s).

The key determinants of an active defense are saturation (how many delivery means can be engaged at one time) and leakage (what percentage get through short of saturation). For example, a Patriot battery with eight launchers (and four missiles per launcher) would be able to engage up to 32 missiles and/or aircraft (the saturation threshold) before reload is needed; if it used two missiles per incoming vehicle, the saturation threshold would be 16 vehicles. If each Patriot missile had a 50 percent probability of killing an opposing ballistic missile, leakage would be 50 percent if only one Patriot were fired per adversary missile, and roughly 25 percent if two Patriots were fired per missile, assuming the kill probabilities were independent of one another.

Beyond saturation and leakage, defenders face other concerns in the challenge and response cycle. A special forces team can more easily destroy a Patriot missile battery than an inaccurate Scud missile can. Adversaries are likely to use combined arms against active defenses, so the United States and its allies need to protect their active defenses. Doing so should let them remove a large fraction—but not all—of the threat.

Avoidance. Force operations can be changed to reduce their vulnerability to adversary CBW threats, even after the United States and its allies develop enhanced defenses. Avoiding contamination is one of the three doctrinal approaches to reducing U.S. vulnerability.[34]

Consider the following analogy. In World War II, the Soviets massed their forces—up to 600 artillery tubes per kilometer—in a breakthrough sector to maximize their ability to penetrate German lines. But in the Cold War, as their vulnerability to U.S. nuclear weapons became clear, the Soviets adjusted their massing factors, eventually settling on artillery densities of about 100 tubes per kilometer in breakthrough sectors. Such a reduced force density complicated breakthrough efforts but prevented a dramatic failure if the opponent targeted the massed forces.

---

[34]Joint Chiefs of Staff, *Joint Doctrine for Operations in Nuclear, Biological, and Chemical (NBC) Environments*, Joint Pub 3-11, July 11, 2000, p. III-6.

To avoid contamination, eight basic adjustments can be made to force operations:

1. *Density reduction.* Forces can be spread out in depth, more reserves can be provided (to replace forward-most forces suffering serious damage), and fires rather than forces can be massed. Density in the rear can be reduced by spreading forces across more bases.

2. *Standoff.* CBW effects can be avoided if forces perform their missions from bases outside the adversary's range.

3. *Dispersal/evacuation.* Forces and personnel, especially those associated with fixed facilities, can be moved away from likely targets (preemptive dispersal/evacuation), and targets already contaminated (remedial dispersal/evacuation).

4. *Relocation.* CBW effects can be reduced or avoided if forces en route to contaminated bases are rerouted to uncontaminated ones.

5. *Avoiding CBW contact.* All forces can try to be indoors during the 30-plus minutes a cloud of aerosolized CBW is passing through. Forces near or in contaminated areas can use reconnaissance to locate and then avoid or decontaminate such areas. Ground forces can maneuver around contaminated terrain.

6. *Threat avoidance.* Ground forces can mount substantial reconnaissance efforts well in front of advancing forces (perhaps 100 km or so) to discover CBW and their delivery systems and destroy them before they are used.

7. *Sequencing operations.* Standoff operations can be used at the start of a conflict to interdict CBW and their delivery means. Once this task is almost finished, forces can enter the theater in greater numbers to complete operations without facing as high a risk.

8. *Quarantine/travel limitations.* Quarantines prevent or limit the spread of contagious BW. Travel limitations imposed during the incubation period of all BW can prevent the appearance that BW were also used elsewhere (i.e., the limitations keep persons exposed at the original site from appearing to have become infected at another location).

Passive Defenses. The final component is passive defenses, which aim to prevent damage from CBW use or to mitigate CBW damage that has occurred. Preventive measures include detection, protective clothing, collective protection shelters, cleanup of residual effects, and consequence management (medical care, personnel replacements, and public information). Figure 2.5 summarizes the current approaches to passive defense, illustrating differences between the CW and BW approaches. The following paragraphs discuss the various elements of the passive defense process.

CW detectors can detect the threat in time to avoid it, but BW detectors cannot currently do so. Thus, passive defense against CW "detects to protect," though CW effects could occur without detection, especially among a civilian population. The persistence of many CW agents also requires reconnaissance and decontamination to finish the job. Passive defense against BW "detects to treat," in part by determining when and by what agent(s) victims have been affected so as to start treatment as soon as possible. Because BW detectors are few and far between, most BW attacks will be detected only by disease surveillance when victims show up at hospitals and clinics.

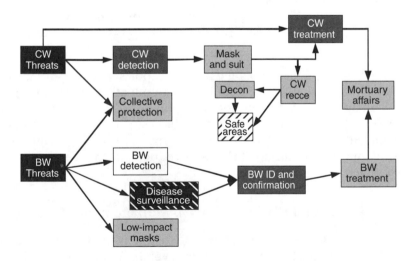

Figure 2.5—Current Passive Defense Efforts

Collective protection keeps CBW out of buildings; it helps those who work indoors avoid exposure. Recent seminars have suggested that it be required in all new construction within potential combat zones. A recent initiative seeks to develop masks that would have less operational degradation than today's CW masks and yet protect against BW (recognizing BW detection limitations) after BW have been used in a theater. CBW medical treatment would be helped by agreements with other governments to provide supplies and specialists, as needed.

Because no single protection suffices, efforts must be combined across them to provide protection packages. Collective protection, for instance, is an ideal defense against CBW for forces that work indoors. If such facilities also have CBW detection, people can be warned to stay indoors when threats loom. By contrast, ground forces in forward units rarely have collective protection against CBW. Masks and suits would be their principal line of defense (plus CBW detection to tell them when to put this equipment on). To avoid overloading their masks and suits, forward units would also maneuver around any contamination they detect by CBW reconnaissance. Reducing the density of forces decreases their value to the adversary as a target and provides reserves to fill holes created by successful adversary attacks. As mentioned earlier, far-forward reconnaissance screens are needed to detect CBW delivery systems and destroy them first.

Although some people working in airfields, ports, and logistics facilities work indoors and can benefit from collective protection, the many people who work outdoors require masks, suits, and CBW detection. In addition, they would be best supported by the use of substantial active defenses—provided by air, missile, naval, and special operations forces (SOF)—to reduce the burden on their passive defenses.

Institutionalization Through Threat Management

The second element of institutionalization is to try to manage asymmetric threats by inhibiting their development in the first place, deterring their use, and mitigating the damage that adversaries seek from them. This element needs to combine prevention, dissuasion, preemption, deterrence, and information operations. While the focus here is on WMD and, specifically, CW and BW as examples of asymmetric threats, in many ways it will be easier for the United States to manage these threats than other asymmetric threats.

Prevention. It is best to prevent asymmetric threats from being developed or, if this is not possible, to get countries to reduce or destroy them. Most prevention strategies are based on arms control; they assume that agreements can be made and enforced to limit or remove certain classes of threats, as was accomplished by the Strategic Arms Limitations Treaty (SALT) and CFE agreements. It is difficult to enforce or verify agreements covering WMD and ballistic missiles even in the case of signatories, however, and many countries are not signatories. The history of Iraqi resistance to WMD-related sanctions suggests that sanctions are no panacea either.

The Bush administration's emphasis on preventing WMD threats runs into obstacles, in particular the "demand" for WMD:

- *Strategic demand.* While the United States has agreed to give up CBW, it still feels it must possess a significant number of nuclear weapons for legitimate national security reasons, including the threat of reprisals. Many other countries appear to feel the same way, though most have focused on CBW because they are less expensive than nuclear weapons. For many of these countries, WMD are the ultimate guarantor of regime survival. They also stand as a symbol of national power while providing some level of coercive power. The United States has not recognized most countries' possession of WMD as legitimate for these purposes.

- *Operational demand.* A number of countries apparently view possession of WMD as an operational military advantage. This is most obvious in countries such as North Korea and Iran, whose inventories of thousands of tons of CBW are difficult to justify for strategic purposes alone.

It is extremely difficult to get countries with these interests to renounce and destroy all their WMD. The United States has not found alternatives for these strategic and operational needs in most cases, making arms control very difficult, especially with regard to countries already possessing these capabilities.

The United States also has arms control concerns on the "supply" side:

- *Moving to "zero" supply.* While the United States has "done away with" all its BW, it still maintains small stocks of many biological agents for defensive development purposes. Recently, it was revealed that even in the United States, the government construed such defensive purposes to include fairly significant biological production efforts to see what a terrorist or other group might be able to produce.

- *CBW breakout.* With BW, most countries prefer to keep primarily small "seed stocks" because of the dangers of storing large quantities and because large quantities can be grown from smaller ones in days or weeks (it takes longer to weaponize the resulting products). Thus, even if a country accepts a "zero" supply, it may be able to produce a substantial wartime capability within months, given adequate expertise and appropriate facilities. Most CBW production facilities have dual civilian and military purposes. Most or all of their production in peacetime might be for civilian purposes, but this could change with little or no warning and few (if any) observables if the decision to prepare for conflict were made.

- *CBW inspections.* Inspections of potential production and storage facilities are a key means for catching countries that are violating WMD restrictions. But since BW production can be done quickly and covertly, it is extremely difficult to catch violators. (Indeed, the UN experience with Iraq after the Persian Gulf War suggests how difficult this process can be.) Moreover, countries are reluctant to enter agreements requiring inspections because they thus open themselves to espionage against key new bioengineering industries. Concerns in this regard have caused the United States to reject the proposed inspections protocol for the Biological and Toxin Weapons Convention.

If all countries were to move to "zero" WMD, a potential adversary could quickly and with little or no notice produce significant quantities of CBW (especially BW) to create an unacceptable coercive or warfighting advantage. Arms control of WMD thus provides only a limited ability to prevent WMD threats. Looking beyond WMD, arms control has even less potential for success with other asymmetric threats such as information warfare and terrorism.

Dissuasion, Intelligence, and the Planning Framework. U.S. efforts to dissuade adversaries from developing or possessing certain asymmetric threats depend on an early understanding of the potential threats, the principles of the challenge and response cycle, and the willingness to invest in capabilities that nullify the threats. If, for example, the United States, anticipating that the smallpox virus has been developed for use against U.S. forces, develops and deploys a smallpox vaccine, few if any adversaries will see utility in further virus development, production, or use. But if the United States does not develop, much less deploy, a vaccine until it has firm evidence that several countries already have weaponized smallpox, it loses the opportunity to prevent this threat, and must instead turn to countering it (which also depends largely on the smallpox vaccine).

Traditional defense planning has been threat-based planning, which copes poorly with asymmetric threats. The defense policy and planning community and the U.S. intelligence community have traditionally required confirming evidence of threats before those threats are included in intelligence estimates.[35] The time it often takes to acquire such confirming evidence means that traditional intelligence estimates lag most adversary capabilities by several years; and they lag adversary asymmetric threats by more years because of a lack of emphasis coupled with the normal difficulties of observing such threats. For example, the production and weaponization of BW agents produces very few of the observables that intelligence agencies usually pursue. Even if BW were observed, the United States would have little information about how they might be used. And since U.S. R&D programs have historically been justified based on such established threats, the United States has tended to seriously lag in the challenge and response cycle associated with asymmetric threats.

---

[35]Ironically, DoD usually ignores the opposite requirement: to confirm that an adversary is *not* developing such a threat before excluding the threat from planning.

The answer for DoD is to try to understand the spectrum of potential asymmetric threats and U.S. vulnerabilities to them.[36] Only then can the United States identify potential threats, search for related adversary developments and strategies, and respond to them. This is the essence of the "capabilities-based planning" proposed in the 2001 QDR. The United States can refine this spectrum of threat alternatives based on cultural and other details of the potential adversary.[37] Planning would then seek to respond to a reasonable threat spectrum.

In essence, the dissuasion strategy element calls on DoD to recognize potential threats and to jump on the leading edge of the challenge and response cycle (rather than being on the trailing edge). It takes the threat spectrum identified in capabilities-based planning, prioritizes it based on threat likelihood and seriousness, and focuses on preventing the top priority threats. To prevent serious threats, the United States must be prepared to invest significantly in developing and fielding appropriate threat counters (such as the smallpox vaccine) well before intelligence can confirm the existence of the threats. The United States must then be prepared to describe the counter to the world in order to convince potential adversaries that the related threat is no longer of use.[38]

Preemption. An alternative to two of the forms of prevention, arms control and dissuasion, is preemption—destroying the threat before it can be used or soon after an early use. Many who heard the 2002 State of the Union Address believe that the United States is threatening North Korea, Iran, and especially Iraq with preemption of their WMD if they do not accept WMD arms control. The U.S. national

---

[36]A focus on known threats leaves the United States open to developing threats. Adversaries would logically start military planning from objectives and develop strategies to achieve them—in part by exploiting perceived U.S. vulnerabilities. Planning of U.S. counters would logically start from a similar perspective. Because the United States may not recognize its own vulnerabilities, this effort must be pursued in part by experts who understand U.S. operations and what can be done to counter them.

[37]One example would be North Korea's interest in special forces and weapons they could use effectively (such as BW), as noted above.

[38]The United States should exercise some care in doing so, as the counter it has developed may, in turn, become the focus of adversary searches for vulnerabilities.

security strategy published in September 2002 included strong support for preemption.

Preemption is akin to attack operations, but would be undertaken as a U.S. offensive action before a war existed with the country in question. Nevertheless, such a preemption would likely begin a war; indeed, North Korea has claimed that President Bush's 2002 State of the Union Address was "little short of declaring a war" on North Korea.[39] In the end, the United States would need to (1) justify to the world that the WMD threat warranted a war, and (2) prove to the world that the preemption largely or entirely removed the WMD threat. To satisfy most countries with regard to the first point, the United States will most likely have to do much more in the way of information operations (many countries do not think WMD threats warrant even President Bush's accusations). And as described earlier, in the discussion on attack operations, the United States will have difficulty entirely destroying the WMD (it will take days or weeks, and may not be possible at all). Moreover, attacks on adversary WMD may well force the adversary into a "use it or lose it" mode in which it responds to preemption with WMD use—for which the United States may then be blamed.

Deterrence. Because protection cannot prevent all losses to asymmetric attacks such as CBW, the United States must deter specific CBW uses.[40] Many equate deterrence with an ability to impose unacceptable damage on the adversary (a reprisal), but the 2001 QDR makes it clear that deterrence by denial (preventing CBW attacks from being effective) is the essence of the new U.S. strategy. Deterrence by denial overlaps very heavily with protection, as discussed above. Still, reprisals do have a role, as do post-conflict sanctions. Because reprisals and sanctions with penalties require that a CBW attack be attributed, the United States needs to prepare plans for achieving attribution in the more difficult cases, such as covert use of BW.[41] And because reprisals and sanctions are escalatory, the United

---

[39]Quoted in David R. Sands, "North Korea Assails 'Axis' Label," *The Washington Times,* February 1, 2002, p. 1.

[40]See Donald H. Rumsfeld, "Toward 21st-Century Deterrence," *Wall Street Journal,* June 27, 2001.

[41]Sometimes attribution is easy. Ballistic missiles, for instance, can usually be traced back to their source (though the country's leadership could claim that they did not

States will need an ability to control escalation, as well as clear knowledge of the price that it and its allies are willing to pay to resolve an adversary's CBW threats.

Inasmuch as no country today is using CBW against the United States, deterrence may be said to be working. The risks of CBW use may far exceed the gains to be achieved, given the probability that those gains can be achieved. But terrorists may feel exempt from such assessments. And if state adversaries begin to feel desperate, their deterrence calculation may change. For instance, if the North Korean regime feared that it was going to be overwhelmed by internal opposition, it might invade South Korea to unify its own citizens; CBW might be an integral component of this invasion. Deterring the desperate from using CBW or other asymmetric threats is hard. The United States can try to prevent desperation through aid, and it can try to convince countries that warfare and CBW use will not solve their problems, or even permit them to achieve what they might want. Potential adversaries should be convinced that the more likely outcome from a CBW attack will be disaster from U.S. reprisals and international economic and political penalties.

Information Operations. Any WMD use would be major international news. Adversaries might seek to ward off penalties, such as international sanctions, in many ways: by denying responsibility, by claiming they are responding to earlier WMD use by the United States or its allies, by depicting themselves as David against the U.S. Goliath, and/or by hindering news coverage of the attack. In using WMD against third countries, they may not even need to pursue these approaches. For example, Iraq's use of CW in its 1980s war with Iran provoked little international response.

---

authorize the launch and will punish the perpetrators). But proving the origins of special forces, terrorist groups, or even cruise missiles is difficult. Without confirmation, the United States may be wary of attacking the wrong party and suffering embarrassment and thus may not act. Adversary deception makes attribution even harder. Given, say, a CBW attack against a U.S. base in the Persian Gulf while Iran is acting militarily against U.S. interests, the United States may well assume Iran did it. But Iraq may actually have done it, hoping the United States would blame and then hurt its archenemy, Iran, in reprisal. The resulting change in the regional balance of power may well be worth the risk for Iraq. Better U.S. capabilities for attribution are needed, and a doctrine of presumed responsibility should be considered, as discussed below.

In peacetime, the United States needs to "prepare the information battlefield." It needs to do so by demonstrating that it and its allies have no plans to use CBW and no deployed capabilities (U.S. CBW has been or is being destroyed), and that U.S. adversaries do have substantial CBW stocks, which they plan to use in wartime and which can inflict a level of damage that easily justifies U.S. and allied protection/defensive efforts. Patriot and weapons systems like Patriot need to be clearly characterized as defenses against serious adversary threats even though some countries' psychological operations (PSYOPs) describe missile defenses as offensive systems.

To justify reprisals, the United States needs to prove that U.S./allied forces were attacked by WMD, convincingly demonstrate the attacker's identity, and show that the United States and its allies have not used CBW and thus are not responsible for whatever CBW casualties occurred in the context of U.S. operations. Otherwise, global non-proliferation efforts will be jeopardized. The United States must clearly describe the damage (especially civilian damage) that resulted from the adversary's CBW use if it is to justify not only the post-conflict sanctions needed to remove the CBW threats, but also the U.S. reprisals—especially if these were carried out with nuclear weapons.

Achieving these objectives in the face of enemy propaganda will be challenging. It will be critical to set international expectations for wartime objectives. Absent clear attribution of adversary CBW attacks, the United States may have to implement a doctrine of "presumed responsibility" that places the burden of proving innocence on regional adversaries with large CBW inventories.

Finally, consequence management requires that the United States be prepared to describe the character and anticipated consequences of a CBW attack, as well as the procedures (e.g., evacuation, medical treatment, and psychological operations to reduce panic) that need to be undertaken after the event.

Internationalization

The United States must also internationalize its response to asymmetric threats, because it will invariably need the support of allies in future conflicts. DoD must prepare itself to fight alongside allies against asymmetric threats, sharing equipment and CONOPs with allies until the latter can meet U.S. standards, which most cannot do yet. Current U.S. efforts offer the opportunity to involve allies in formulating CONOPs and equipment R&D, thereby involving them and potentially reducing U.S. costs. Unfortunately, many U.S. efforts to develop counters to asymmetric threats are pursued on a "U.S.-only" basis, constraining and often directly thwarting U.S. efforts to internationalize the results. This is particularly true with regard to the development of new technologies.

Internationalizing the U.S. approach to CBW threats involves the following five elements:

Understanding the Threat. The United States will need to help its allies understand the CBW threat. Much U.S. information is acquired through technological means; the United States can gain much from the human intelligence (HUMINT) and cultural awareness of its regional allies (especially relative to terrorism). The United States must exchange this information with allies systematically and comprehensively, taking into account all concerns about sensitive sources and recognizing the potential unreliability of such information. The United States and its allies might disagree on the relative likelihood of elements across the threat spectrum, but the spectrum itself should be an easy basis for the consensus needed to establish a common background for strategy and planning.

Synchronizing Operations. Most allies lag the United States in developing CONOPS for CBW threats. Once allies appreciate such threats, they generally welcome U.S. ideas for reducing force vulnerabilities, especially when those vulnerabilities can be reduced merely by adjusting CONOPS. Because U.S. allies lack many of the protections against CBW threats, the United States needs to field phased sets of CONOPs that cover situations from minimal protection through well-developed protection. For example, the first thing all airfield workers should do after a CW attack is get indoors and turn off the ventilation systems that could bring in contamination from the outside. Those who have protective clothing should then put it on; those without protective clothing may not be as well protected, but they will survive better indoors than out.

It is important to set operational norms. For example, in a serious CBW environment, forces ought not be committed to frontline roles without protective clothing. Having this restriction mandated in combined planning would help allies see that acquiring such clothing is a priority if their forces are to be included in desired operations.

Sharing Protections. The United States leads most of its allies in developing protections against CBW. Some of these have been made available to U.S. allies,[42] but not all of them have, so most allies will have inferior protection or none at all. The United States should share as many of its protections as it can in order to facilitate combined operations and promote the use of similar tactics, techniques, and procedures. DoD also needs to appreciate how leaving allies less well protected may lead to sharp criticism of the United States in the foreign media. Until allied protection comes up to U.S. levels, the United States should balance the protections—e.g., offer active defenses (such as Patriot) for key allied facilities that have little or no passive defenses. Such acts will strengthen combined efforts and help to secure allied involvement in difficult conflicts.

Coordinating Destruction. In targeting adversary CBW supplies, there must be clear coordination of U.S. and allied military planning to reduce duplicated effort, avoid collateral damage, and minimize contamination that would impair future maneuver or other operations. Such concerns would be particularly important if U.S. nuclear weapons were used in attack operations. Because most CBW are stored underground, U.S. nuclear attacks on them would rely on ground bursts, which generate fallout. Cleaning up this fallout would be expensive and time consuming. Thus, nuclear weapons should be used against targets only where fallout can be avoided or under wind conditions that would deposit fallout in less sensitive areas. Also, the destruction of BW could lead to a spread of contaminants that would require U.S. and allied efforts to quarantine or control travel in the

---

[42]There are various reasons for not sharing protections with allies: (1) other allies may have helped to develop a protection and may thus be in a position to disapprove its release; (2) a protection that is understood can be countered, so sharing information about a protection with allies makes it more likely that adversaries will find out the details; and (3) some U.S. protections are just too expensive for allies to purchase in sufficient quantity.

affected areas. Finally, because special forces are often the best way to find and destroy CBW stocks and delivery systems, counter-CBW roles for allied special forces need to be developed.

Cooperating to Prevent and Dissuade. Given how hard it is to prevent the development of CBW and their delivery systems, such a task is almost always best performed as a coalition. It is hard for the United States, even as a superpower, to dictate terms to other countries.

Cooperative prevention requires the United States to share with allies a common understanding of CBW threats and their implications. The United States then needs to reach agreements on appropriate objectives for prevention efforts: Can CBW development be completely stopped? If not, what limits are achievable? What must be done to achieve these objectives? How well can prospective adversaries hide or otherwise conceal their CBW efforts? In this regard, the U.S. experience with Iraq is discouraging and suggests that it will be difficult to prevent CBW developments without some degree of cooperation from the countries under suspicion.

CONCLUSIONS

Adversaries are unquestionably pursuing asymmetric threats to counter U.S. and allied military power. U.S. and allied responses are limited by adversary efforts to conceal these developments, forcing the United States to deal with a very uncertain threat spectrum. Dealing with such threats requires an approach having many dimensions—not least of which are forces and CONOPs to sustain military operations against asymmetric threats and to prevent or counter the surprise and operational disruptions that adversaries will seek to achieve.

Because asymmetric threats are diverse and ever-changing, challenge and response cycle analysis seeks to keep U.S. responses ahead of the cycle's curve. Finally, key capabilities, such as the ability to attribute attacks and control escalation, need more attention from U.S. planners.

Chapter Three

WHAT INFORMATION ARCHITECTURE FOR DEFENSE?

Martin C. Libicki

We live amidst an information revolution, which is to say, a revolution in the capabilities of information technologies and infrastructures. The quality and quantity of the information we receive have greatly increased, but for information to be truly useful, it must improve the quality of our decisions, which, in turn, are judged by the quality of the resultant actions.

Two-thirds of all personal computers and almost all networks and databases are used for business, not recreation. Plausibly, therefore, most decisions that information technology supposedly improves are those made in an organizational context. They result, one way or another, from interactions among people and their machines. As such, the quality of the decisions an organization makes is increasingly related to how it constructs its information systems. These architectures are often faithful reflections of the tacit assumptions about power and purpose held in these organizations. Architecture and organization are linked. An organization's tendencies shape its architecture; its architecture, in turn, helps shape its culture.

The theme of this chapter is how organizations respond to the opportunities and challenges that come from giving everyone access to rapidly increasing amounts of information and corresponding tools, such as analysis and communications. Should sensors be few and commanded from up high, or many and commanded from the trenches? Should information be pushed to the user based on what is deemed necessary, or should users be able to pull information and subscribe to data-flows they feel they need? Can information be arranged and rearranged by users to best fit their intuitive grasp of the matter, or does one display fit all? Will bandwidth constraints limit the information and services users can access, or can a robust menu of alternatives be employed to expand or work around such obstacles? How much should be invested in providing the capability needed to discover information in realms ostensibly far from one's learned domains? Must security be enhanced by limiting what warfighters can know of the battlespace? Should interoperability be good enough to permit users to seamlessly skip from one domain to another, plucking what they need from what they can see? Will users come to feel they own their tools?

WHAT IS ARCHITECTURE?

A common-sense definition of architecture holds it to be the relationship of a system's components to each other. But since people are elements of almost all information architectures, architectures involve human participation. Like any policy structure, architecture reflects power relationships. Unlike many policies, though, it can alter the social underpinnings of power relationships.

The subject of information architecture has more of a future than a past. All organizations have an information architecture, but before there were information systems, the fundamental principles of an organization's information architecture directly reflected (1) command relationships and related functional responsibilities, (2) geographic distribution of personnel, (3) the distribution of clearance levels and related physical access privileges (e.g., keys, safes), and (4) personal relationships. In other words, information architectures were a direct reflection of management and institutional relationships and could be almost completely studied in that context.

The development of the telephone and telegraph altered this formulation by removing some of the geographical impediments to communications. Complex machinery with its sensors and controls possessed its own information architecture, but it was generally internal to the machine itself. Rarely were these machines instruments of an organizationwide architecture. Prior to World War II, the percentage of all workers with phones at their workstations was low, and the ability to control machinery remotely was negligible.

The entry of automated data processing systems in the 1950s created specific work flow paths for certain types of information—e.g., payroll processing and inventory management. Most people accessed computers through special-purpose terminals that limited their interactions to predetermined processes. General access to corporate information was rare except within top levels of management. Indeed, for the first few decades of their existence, computers tended to have a centralizing effect in that they permitted management to harvest much more information about its enterprise.

The development of recognizably modern information architectures began not much more than a quarter-century ago, when sophisticated measurement and control systems were coupled with workstations cheap and robust enough to proliferate outside computer rooms. Widespread personal computing in the workplace is no more than 20 years old; networking, no more than 15; and general Internet connectivity, no more than 10 (universities being the primary exception). The advent of pervasive information systems has so greatly changed the quantity of information available to people that it has changed the quality and hence the nature of the architectures.

Organizational information architectures are still evolving. Office workers in most modern Western organizations have workstations with intranet and Internet access. Most production and paperwork processes are monitored throughout their journey and have their minute-by-minute information logged. These data are widely, but not necessarily generally, available. Corporate America is in the midst of switching from phone and paper to the Internet as its primary link to suppliers, collaborators, and customers.

The next decade or two should bring further complexity and richness in organizational architectures, thanks to several trends:

- Access to both the Internet and organizational intranets through initially low-bandwidth and always-on small-screen mobile devices, cell phones, or cousins of the Palm Pilot VII™.

- Greater systemization of institutional knowledge bases to the point where such knowledge can be analyzed with data mining and other forms of logic processing.

- The growing ubiquity of small sensors and smart radio-frequency bar-coding both in process industries and in public,[1] and in quasi-public settings (e.g., hospitals, schools).

- Further increases in the extraction of usable knowledge about customers and the overall environment from external sources both public (the Web) and quasi-public.

- The granting to suppliers, collaborators, and customers of deeper access into one another's corporate knowledge bases.

The Global Information Grid

DoD's interest in information architecture is clear. Joint Vision 2010 and Joint Vision 2020 are paeans to the importance of information superiority to military superiority. Many of the technologies used to construct information systems were developed by or for DoD. For a long time to come, DoD's information architecture will revolve around mobile devices and sensors of the sort that corporate America is only starting to see. But the levels of automation within DoD vary greatly, partially because capital turnover is slower there; large parts of DoD are more digitized than is corporate America. It is a cliché to note that DoD has no information problems not shared by private enterprise, but there are real differences between the two.

DoD is beginning to contemplate an enterprise architecture for itself. It already has bits and pieces under way within service systems (e.g., the Army's nascent Force XXI initiative and the Navy's IT 21 project), joint internets (NIPRnet for unclassified work, SIPRnet for secret material), and its software suites, the Global Command and Control System (GCCS) and the Global Combat Support System (GCSS). The term *global information grid* (GIG) is often used to describe the anticipated agglomeration, variously known as the global grid (the 1992 JASON study), the system of systems (former Joint Chiefs of Staff [JCS] Vice Chairman Admiral William Owens, but also former

---

[1]The events of September 11, 2001, are likely to sharply increase the use of identification cards, such as proximity cards, for automatic logging of entries and exits.

Army Chief of Staff General Gordon Sullivan), and the battlespace infosphere (U.S. Air Force scientific advisory board).

How far along is DoD on its architecture? DoD typically divides architecture into *systems architecture* (what is "wired" to what), *technical architecture* (how interfaces are defined), and *operational architecture* (how data flow to carry out missions). Systems architecture exists, by definition, at every point, but whether it accords to its specifications is a different matter. In practical terms, the shortfall between how DoD's networks are configured and how they ought to be configured given equipment constraints is probably modest. Technical architecture also exists, but the persistence of legacy systems means that the actual state of interoperability falls far short of where it would be if all systems complied with the joint technical architecture, which, itself, covers only a small fraction of what it ultimately has to.

Operational architecture remains largely undocumented, although there are calls to make it more formal. Systems architects could determine where to make investments if they knew where the data were supposed to flow; technical architects could determine where interoperability was most critical if they understood where information had to be exchanged and understood. So what is the hang-up? Operational architecture sounds easy, but it is about such elemental issues as who talks to whom and which processes need what data— which is to say that it is about power, and power is extremely difficult to negotiate and codify in peacetime. Moreover, the nagging feeling persists that, to echo Von Moltke, no operational architecture will survive contact with the enemy. War generates surprise. Many systems fail in combat; others succeed brilliantly but are too few in number.

Yet all three forms of architecture somehow fail, even in combination, to capture the essence of architecture, the constructs under which people exchange information. Physical architects enter school believing that architecture deals with the arrangement of structural elements; they leave school understanding that its subject is really the arrangement of spaces within which people interact. And so it is for information architecture.

Need There Be Architecture?

One approach to architecture posits a hierarchy of tasks.[2] Physical architecture begins with a statement of requirements (the "bubble drawings" that rough out room arrangements) that is successively refined through architectural drawings, construction plans, subcontracting processes, and bills of material. Information systems are, likewise, successively broken down from task requirements to implementation along three parallel paths: data structures, algorithms, and communications networks.

The Internet was built on entirely different principles, even if few people remember that its first purpose was to share computing power too expensive to duplicate everywhere. Data structures (except for addressing), algorithms, and even routing are now entirely beside the point. The Internet does not *do* anything. It is an infrastructure used by a great many people to do a wide variety of things. The Internet does have a straightforward and explicit set of rules that govern how networks become members and format their message envelopes. Otherwise, if architecture is defined as who-can-say-what-to-whom, the Internet has very little of it.

The Internet works famously, so why bother with architecture at all? Why not simply let everyone make information according to their abilities and take it according to their needs? Why limit the resulting conversation? There are several reasons.

First, the Internet is primarily a transport mechanism and hence only part of an information system. To accomplish specific tasks within specified parameters, organizations have had to impose an information architecture on top of it. Even the research community, where the ideals of the Internet are most fully realized, has architectural elements embedded into its culture, such as how material is published.

Second, the Internet's survival for most of its life has been based on shared norms and the kindness of strangers. As it has opened itself up to literally everyone over the past five years, it has faced tough

---

[2]J. A. Zachman, "A Framework for Information Systems Architecture," reprinted in *IBM Systems Journal*, Vol. 38, No. 2, 1999, pp. 454–470 (originally printed in 1987).

problems in privacy protection, intellectual property rights, spamming, and poor security.

Third, even if the principles of the Internet and World Wide Web are the goal of an information infrastructure, such ideals must be made manifest in a world where truly open organizations are rare. Yet not every organization has an information architecture per se. RAND does not have one for its research, although it does have a culture. RAND's ability to get work done depends on researchers acquiring information from outside the corporation and analyzing it using tools they know very well. In effect, RAND rides on the information architectures of its clients and the research community.

Architecture Follows Culture?

How cultural norms affect people's participation in the information era is a big and well-studied subject. But it helps to point out some general questions that suggest there are large variations in how different cultures use information. Intuition says that cultural factors can help or hurt an organization's use of information systems—especially people's willingness to trade information, seek out potential disquieting knowledge, undertake honest analysis, and base decisions on the results of that analysis. Other factors, such as the tradeoff between horizontal and vertical flows, between public and private credibility, and between oral and written communication, ought to affect information system design as well. Part of what DoD (or any organization) ought to do in developing an enterprisewide information architecture is to examine its own cultural assumptions and the ways in which they may feed into architectural decisions—a process akin to warmup exercises before undertaking a long run. To understand any given culture, one should ask

- Are people more likely to hoard and then trade information than they are to share information based on trust?[3]

- Are people more likely to believe what they get from public sources or what their friends tell them?

---

[3]Frank Fukuyama argues that cultures can be characterized as high- or low-trust. See his *Trust: The Social Virtues and the Creation of Prosperity*, The Free Press, New York, 1995.

- Does more information flow vertically or horizontally? Is vertical collaboration accorded the same weight as horizontal coordination?

- Is information passed via the written word or the spoken word? Which form conveys more authority, granted that this often depends on personal preference?

- Is there a bias for acting and letting the facts flow from the results rather than undertaking analysis before acting?

- Must decisions be justified by facts and arguments, or do authority, experience, and/or charisma suffice? To what extent are formal credentials taken seriously as a source of status?

- Will people seek out knowledge even when it may contradict their earlier judgments—especially publicly offered judgments?

Underlying the premise that information systems based on sound architectures can help an organization is the requirement that an organization be sound and coherently organized. A few test questions to make this determination might be:

- Is the organization in the right line of work? Is it looking to the right customers and solving the right problems? (Solving the wrong problems more efficiently is of only modest benefit.)

- Is the achievement of organizational goals accorded a higher priority than the achievement of any one faction's goals? of personal goals?

- Is honesty accorded respect? Can the organization accept bad news without shooting the messenger or inventing fantasies?

If the answers are affirmative, information and thus a good information architecture can help. DoD seems to score above average on all three questions.

DoD as an Institution in Its Own League

DoD differs from other institutions, so many of the issues raised concerning its architecture may be of less consequence to those other institutions. The differences are many.

DoD is a hierarchical organization that relies on command-and-control (especially in the field) irrespective of the distribution of knowledge, and it is likely to remain so regardless of how technology evolves. Warfighting puts people at personal risk, giving individuals a potential motivation that can be at great odds with their organization's motivations. Militaries are built to engage in contests and to serve the national will. Most other organizations, by contrast, respond to whoever will pay them; most customers are individuals. DoD has multiple layers, each with its own assets and requirements, and each strongly biased toward owning its own information sources.

Because DoD needs to know the whats and wherefores of uncooperative foes, information collection is expensive and problematic. Sensors are expensive, deception is the norm, and militaries are obsessed about information security. Businesspeople at least can rely on the force of law to inhibit mischief by competitors; militaries work in an anarchic milieu and thus cannot.

Because DoD works outdoors, it relies heavily on radio-frequency links, often established in austere regions (e.g., at sea). Nodes are often at risk of destruction; communications is ever subject to jamming, so transmission and reception equipment must be hardened. Bandwidth constraints pinch DoD more than they typically pinch commercial enterprises. Combat communications are often urgent, and many of those who have most need of information have limited attention to give to what they see or even hear.

Finally, DoD cannot rely on outside sources for all of its education or institutional learning. Much of what it teaches has no outside counterpart, at least in this country, and often not anywhere. Bill Joy, the cofounder of Sun Microsystems, once observed that regardless of who you are, most of the smart people in the world work for someone else. DoD can occasionally be the exception to this observation. Most of the people who best understand, say, stealth technologies work for DoD either directly or indirectly. This also means that if the state of the art in areas of DoD's interest is to advance, those areas must be deliberately resourced by DoD itself.

Perhaps the most interesting facet of DoD's unique requirements is its imperative to optimize not efficiency, but adaptability—particularly now, when it is so difficult to make reliable statements about where or against which foe the U.S. military will have to practice its craft next (as September 11 and its aftermath have once again proven). To illustrate this point, Table 3.1 categorizes warfare according to whether mass or precision matters more, and whether the United States gets to fight from standoff range or has to move close in. Every square holds a different form of war, each of which demands different features from DoD's architecture.

For the historic, conventional type of warfare—close-in mass operations—commanders need data from the field to build and share almost-instant situational awareness. This information fosters command dominance, which, in turn, permits its possessors to execute decisions faster than enemies can react. The military, however, requires an even larger quantity of *internal* data to exercise its core competence of conducting highly complex maneuvers. Carrier battle groups, theaterwide air tasking orders, and maneuvers in corps are all examples of how the U.S. military can orchestrate the actions of 10,000 to 100,000 individuals better than any other entity can.

For strategic warfare—standoff mass operations—commanders need a more analytic and synoptic understanding of the enemy's pressure points. One method is to learn how the enemy's economy, political structure, and infrastructures (both military and civilian) are wired so as to be able to identify the nodes. This invariably becomes a large modeling exercise.

For hyperwar (DoD's preferred mode), the most pressing problem is to locate, identify, and track mobile targets so that they may be struck quickly from afar. The important qualities are the ability to scan large battlespaces, sift the few interesting data points from the mass of background, sort the targets of potential by priority, and strike those targets while they are still visible. This search for in-

Table 3.1

A Two-by-Two Categorization of War

	Close In	Standoff
Mass	Conventional conflict	Strategic conflict
Precision	Mud warfare	Hyperwar

formation is akin to finding the classic needle in a haystack—or, because targets continually move, finding the snakes among the worms.

Finally, if despite everything DoD is stuck in mud warfare and working within dense milieus, it needs a precise understanding of its environment so that it can distinguish the malevolent from the irrelevant. A precise situational awareness also permits threat patterns to be perceived.

These are, admittedly, gross generalizations. But they illustrate the wide range of functions that DoD's information architecture must satisfy. They also illustrate the potential folly of building the GIG architecture from the top down. If the U.S. military cannot be sure of when, who, or how it will fight, should not its architecture be optimized less for any specific scenario and more to accommodate the fluidity of a real-time, high-density, rapidly changing and restructuring world?

For instance, is the GIG to be understood as a command, control, and communications (C3) system that ties people to each other (and, incidentally, to some interesting services) or as an intelligence, surveillance, and reconnaissance (ISR) system that uses communications to accommodate a distributed workforce? The two facets ought to be two sides of one coin, but the cultures they support differ greatly. Communicators like to share information; intelligence types tend to hoard it. Communicators want to know who talks to whom; intelligence types want to know what information is fused with what.

Who is the ultimate customer: the White House or the foxhole? The intelligence community has been facing this question since the Cold War ended. Initially, both urgency and the scarcity of good information oriented intelligence to the U.S. president. Now, with the Cold War over and open-source information more available, intelligence oriented to the foxhole makes more sense—but how much, and how fast? This question is felt with great keenness in peace operations. Is the primary point of military operations to generate an outcome consistent with U.S. interests or to serve clients (e.g., those who live in an erstwhile war-torn land)? The answer sets up a broader question: Should such a system be designed to spread information down and around or to filter it up?

The issue of who owns information raises command and control (C2) issues. Ownership duly and dually implies both responsibility and control. But is the responsibility to collect information logically connected to the right to control not only who gets to see the information, but also the form in which it is released? That case is hard to make in the age of the Internet and the Web. It would seem that information should be generally releasable, subject to no-more-than-necessary security constraints, broadly accessible, and formatted in the most interoperable way—which often means with the least amount of unnecessary processing. But can potential users counter the collectors' argument that data cannot possibly be released to the rest of the world without being subjected to thorough and time-consuming analysis?

ELEMENTS OF ARCHITECTURE

In this context, eight categories can be used to describe architectural issues related to information: (1) collection, (2) access, (3) presentation, (4) networking, (5) knowledge maintenance and management, (6) security, (7) interoperability, and (8) integration. Each is discussed in turn.

Collection

How an organization gathers data depends on technical considerations as well as its judgment about what kind of information is needed by whom.

What, for instance, should be the mix between hunters and gatherers? Hunters define information requirements and then collect in accordance with them; their questions are specific, and their tools are often focused and may be used intermittently. Gatherers collect large amounts of data and then thresh through what they have for anomalies, telltale changes, and other interesting tidbits; their tools tend to be general and used continuously. The intelligence community has both types of users. The imagery business tends to be filled with hunters, with tools that are continuously busy (e.g., reconnaissance satellites) and tools that are employed only in discrete missions (such as U-2s). The signals intelligence business tends to be filled with gatherers.

Is information to go up or down the logic tree—the oft-cited path connecting data to information to knowledge, understanding, and wisdom? Some use induction, starting with examples and winding up with generalizations. Others prefer deduction: they start with rules and then exploit them to develop differentiations. The first approach is good for figuring out the enemy's doctrine, the second for distinguishing an enemy from a bystander. Inductors start their post-processing late in the game; deductors early.

What integration metaphor should be exploited? One approach (and many may be needed) is scan-sift-and-sort, which is particularly useful for engaging mobile targets amidst clutter. Another challenge is building a synoptic picture from the coordination of distributed sensors. Managing by maintaining parameters (e.g., the odds that a district is secure, the likelihood that the enemy will do X under stress) requires that new facts be consistently integrated into old equations.

How are data validated and reconciled? The Army says the village has 100 enemy holed up, the Marines say 50—which number goes in the database? What criteria are used to point users toward one or another estimate? What metrics are used to label facts as being of greater or lesser validity?

For DoD, many of these questions are arising as its GIG becomes more deeply networked and more inundated by sensor data. The greater the bit flow of any one phenomenology, the greater the logic to move from hunting to gathering successively lower-grade ores simply because it can be done. Many of the more intractable problems of warfighting, such as tracking targets in clutter or mobile emitters, are problems of gathering. Issues of data integration and validation arise as the volume of data grows apace and such techniques as cue-filter-pinpointing, automatic sensor coordination, systematic parameter maintenance, and even the computer-aided marriage of scattered knowns and needs all become more feasible.

For DoD, data collection is increasingly an issue of sensor deployment and management rather than, say, human reconnaissance.

For instance, should sensor systems be designed to collect broad knowledge and to cue weapons that then find the target, or should sensors collect knowledge precise and timely enough to guide weapons to their target? As Chapter Four, "Incorporating Information

Technology in Defense Planning," suggests, the choice between man-guided, seeker-guided, and point-guided weapons is not simply one of engineering, but of empowerment as well. A third party given man-guided weapons has to be operationally competent to make use of them. Given seeker-guided weapons, the third party can do what it wants. With point-guided weapons, it will have to depend on those who supply the points and tracks, so if that "who" is the DoD, the United States retains a great deal of leverage over how the weapons are used.

A parallel question is whether DoD should lean toward using a few expensive sensors (such as billion-dollar spy satellites, Aegis radar, Joint Surveillance Target Attack Radar System [JSTARS]) or many cheaper ones (such as micro unmanned aerial vehicles [UAVs]) or disposable unattended ground sensors. If expensive sensors are used, there will be contention over who gets to use their capacity. A system of cheaper but more numerous sensors lets all operators have their own capability, but what happens when there are more sensors than humans can manage? How far can and should the sensors be designed to manage themselves?

Finally, should sensors output (1) raw data, (2) data cleaned up by some automatic and semi-automatic processing, (3) only data that are completely exploited and have passed quality-control tests, or (4) only data that have been fully fused with all other comparable data? How close should the coupling be among sensors, sensor data storage, sensor post-processing, and sensor data display?

Access

Over the last half-century, management attitudes on information have shifted from need-to-know to right-to-know. Many companies, for instance, have found labor unions to be more tractable if workers are shown detailed information on the company's financial performance. As a general rule, the military has lagged on this issue, although DoD is gradually giving people more access to its information both internally and externally.[4] For example, the end of the Cold War reduced many classification levels to Secret.

Many reasons have been offered for limiting information. Two of them—the cost of making legacy systems interoperable, and limited bandwidth—have technical solutions. The security argument is a shibboleth to the extent that almost all career military personnel are cleared to the Secret level; *operational* information is rarely classified higher unless sources and methods are involved, but they can usually be scrubbed out of the data.

That leaves one real reason for restricting information: it might be distracting. DoD has gotten great mileage out of warfighters with modest education by carefully developing a hierarchy of task structures to which operators are closely trained. Information requirements are then carefully generated for these tasks. Unfortunately, the narrow task specialization that pervades the national security establishment (only 1 percent of all the intelligence community's workers are all-source analysts) is increasingly at odds with how information is handled elsewhere: workers are given an increasingly broad view of their task within an organizational context.

Another possible reason is that warfighters would not risk their lives in dire circumstances if they understood how bad things really were. Such a logic may apply to other militaries, but the U.S. military takes public pride in the competence and professionalism of its warfighters, who expect to be told the truth and take responsibility for keeping focused.

What gets shared beyond DoD? Many of the arguments made against sharing data with coalition partners follow the logic above. For long-time partners, such as Britain and Canada, this argument rings hollow. For those who are made coalition members for political rather than military reasons (e.g., Syria in the Gulf War), a certain hypocrisy that separates the promise to share from the actual information supplied can be expected. How much of DoD's information should

---

[4]After September 11, 2001, however, public access to civilian information appears to have been restricted because of the fear that terrorists could use it to prepare attacks; see Ariana Eunjung Cha, "Risks Prompt U.S. to Limit Access to Data," *The Washington Post*, February 24, 2002, p. A1.

be publicly accessible? In the last five years, DoD Websites criticized as too revealing have cut back. "Guarded openness"[5] may permit U.S. forces to acquire the cooperation and facilitate the coordination of third parties, or at least to get their input for intelligent preparation of the battlefield.

Presentation

In many cases, information flows and needs cannot be matched unless they are formatted in some standard way, which, in turn, means they must be encapsulated and categorized according to some semantic and cognitive structure.

The choices information providers face regarding how to forward information to the field are rich ones. E-mail, pop-up alerts, and monitors are examples of push processes; Websites and query capabilities typify pull processes. A mix of the two types of processes is possible. In a guided tour, for instance, a broad request yields a linked set of answers, elements of which can be invoked based either on the user's interests or on abilities made explicit by the user. A more sophisticated version would infer what the user wishes to know and push selected information (both stored content and news as it develops) based on the specific inquiry, past inquiries, and perceived user habits.

Information architects endlessly repeat the mantra "the right information to the right person at the right time," also adding "with the right presentation and the right security." But who is to judge what is "right"? Some analysts believe that military missions and tasks can be decomposed so that the right information pops up as needed, much like an automobile navigation system builds trip routes and flashes left or right arrows as key intersections are encountered. Users of good operating systems, such as that of the Palm Pilot™, may delight in having the system bring up the right menu of choices whenever one or another action has taken place. There is an ele-

---

[5]Carefully vetting what is released, but then releasing it without restrictions. See John Arquilla and David Ronfeldt, "Information, Power, and Grand Strategy: In Athena's Camp—Section 2," in John Arquilla and David Ronfeldt (eds.), *In Athena's Camp: Preparing for Conflict in the Information Age*, MR-880-OSLVRC, RAND, 1997, pp. 413–438.

gance to good design; this is why people hire architects and try not to draw their own maps.[6] It helps eliminate clutter; it also permits information providers to draw out implicit narratives from media, such as maps, which only the skilled can navigate. Pushing information at users relieves them from having to make explicit choices. It can be valuable when users are overworked or under stress.

To the user, *how* information is presented is a critical adjunct to what information is presented. Simply by highlighting certain information or certain relationships among data elements, a person or program can push everything else to the cognitive background. In a complex environment, data that are not highlighted might as well not exist. But who decides? "Shared situational awareness" is one thing if it means that everyone can access the same information. But should there be a standard method of presentation? If there is, how easy should it be for users to redraw their own map, so to speak, if they can even do so at all? The Common Operational Picture (a GCCS application that indicates the presence of major units on a battlefield map that everyone shares), for instance, permits certain predesignated layers of information to be highlighted or hidden, but it provides no macro language that lets intelligent users (or their information aides) manipulate data fields of unique importance to them. People do, after all, see information in different ways. Having everyone see everything alike has two advantages: it promotes efficiency in conducting complex operations, and it minimizes complaints such as "How come I can't see this?"

But what if the problem is not complex coordination but complex recognition, as may be the case for modern warfare against an elusive and well-hidden foe? Users might usefully be allowed to peer into the weeds in multiple ways on the theory that it may take only one person's burst of insight to find the elusive foe or make everything fall in place for everyone else. Free play exploits the vast differences in how people perceive the world. Making everyone look at the world the same way may not hurt those whose perspective is sufficiently near the official norm, but it may reduce the contribution of those who think differently. Even if people thought similarly, there

---

[6]See Edward R. Tufte's *Envisioning Information*, Graphics Press, Cheshire, CT, 1990; *The Visual Display of Quantitative Information*, Graphics Press, 1992; or *Visual Explanations: Images and Quantities, Evidence and Narratives*, Graphics Press, 1997.

might still be a point to having them learn how to manipulate their perspective on their own until the world comes into focus. They may learn more about perception and pattern recognition for having discovered as much on their own. In an era of information overload, simply going through the exercise of determining which few features of an environment together tell the story is a valuable lesson in recognizing the essential. So presentation, like its close cousin access, also depends on who decides.

Networking

Access often depends on how much bandwidth is available. In a world of thin 9,600 bits-per-second (bps) lines (the top speed of the new Single Channel Ground and Airborne Radio System [SINCGARS]), warfighters can demand neither imagery nor access to databases. Since compression of one sort or another has to take place at the information source, or at least at the last cache before the fiber ends, the prejudice against user choice is easy to explain away as a technology-driven requirement. But is it?

Mobile and fixed access to the Internet differ sharply. Europeans have taken to mobile phones faster than Americans, in part because they settled on digital cellular standards earlier. The race is on to serve them Internet and Web access through small phone screens, although Palm VII™-like devices may be part of the mix. Because cell phones know where they are (to within roughly 300 m), investors are excited about tailoring content to location (pass an ice cream shop, and a pop-up message about new flavors appears on the screen). Americans, by contrast, are more likely to surf while sitting still; they have more screen space and bandwidth to match. Content cannot be presented the same way to both the desktop and the handset. Presentation for the latter must be more parsimonious, favoring text over images, sound over sight, and voice over tactile input.

The differences in information accessibility for mobile versus immobile users are critical for the U.S. Army. Division XXI, the Army's digital initiative, provides a rich array of information devices from the corps headquarters down to the battalion tactical operations center (TOC) using million bit-per-second connectivity for systems that support intelligence, maneuver control, and fire support.

Beyond the battalion TOC, however, soldiers are linked, at best, at 9,600 bps via an Appliqué—a limited terminal to which some information is pushed and from which certain numbered queries can be sent. If the ability to wield power is enhanced both by being close to the field and by being able to access information, the digitized Army appears to pivot around the battalion TOC commander, who alone has a good enough view of both worlds. Is this what the Army really wants? Is this necessarily the optimal configuration for the sorts of wars the Army is likely to fight? Urban combat, peace operations, and new theories of swarming[7] may demand that control be exercised by company commanders, one full level down. The Marines pivot around the squad and the platoon. Will reifying Army doctrine in the hard-to-change hardware and software of its information systems promote jointness or the ability to carry out combined operations with overseas counterparts?

Perhaps the issue is less bandwidth than how bandwidth scarcity is managed, for there are choices other than limiting the functionality of terminals to fit bandwidth limitations. Appliqués could be full-fledged clients that accommodate reduced bandwidth by squeezing information flows, transmitting images with less clarity, less color, less detail, and/or fewer updates, or even abjuring imagery in favor of symbolic transmission. Or, each radio's capability could be enhanced but bandwidth managed explicitly as a battalion-shared resource, so as to use more bandwidth for each radio but employ explicit contention mechanisms to allocate spectrum.

A similar architectural issue comes from the breakdown between the field headquarters and those back home who have copious access to information. If the major source of battlefield information comes from space, and space information comes in gigabytes, will there be an inevitable shift from the field toward those with sufficient access to see everything coming in from space? How will the nascent shift from space to field-supported UAVs affect this balance?

Traditionally, C2 was the metric for distributing bandwidth. Fat pipes connected the commanders in chief (CINCs) with the United States, and these pipes grew successively, if inconsistently, smaller as they

---

[7]For background, see Sean Edwards, *Swarming on the Battlefield: Past, Present, and Future*, MR-1100-OSD, RAND, 1999.

skittered down the ranks. Actual commands or reports are unlikely to overfill the pipes as much as imagery will, but the demand for imagery is not very correlated with rank. The current thinking tries to take operational architecture (who says what and how much to whom) as its foundation, but the calculations have proven vexing. The STU-3 (a telephone handset for making secure phone calls) was said to be the great weapon of Desert Storm because it allowed direct contact between the Pentagon and the field, and, in many ways, circumvented traditional routes of influence. There may be understandable reluctance to accord a patched-up set of relationships equal status with wiring diagrams on paper.

Knowledge Maintenance and Management

Knowledge tells one how to carry out tasks. But in a military context, knowledge also includes broad estimates of reality (e.g., this village is or is not safe) and operational rules of thumb (e.g., a village with few young men visible is generally a hostile one). In deciding how analytic judgments (such as how safe the village is) are to be maintained, one must decide who will have authority over what variables. If automatic processes rather than people maintain judgments, the choice of which variables are fed in and how becomes important.

In any complex environment, there will be many newly discovered facts, including new events, and parameters that change to reflect those facts. The methods used to feed new facts to parameters (and the algorithms by which new facts are ingested) are thus important. Sensors, for instance, may pick up new jeep tracks, a finding that could be used to update the odds that an adversary is working in the area, the specific equipment the adversary has, perhaps how much equipment, and maybe even the adversary's doctrine for using it (why were no efforts made to hide the tracks?).

Many computer specialists are starting to think about how parameters may be maintained by software agents—i.e., processing components passed among machines to perform functions for their owners. To use a biological metaphor, an agent (for a parameter: in this case, evidence of a country's interest in nuclear weapons) may wander among nodes that have receptors for various kinds of events—say, the hiring of engineers with nuclear specialties, market reports of certain chemical sales, and suspicious travel patterns. A match might cause the agent to emit signals that stimulate other agents to activity (to review where the country is sending its graduate students) or lull others into relative inactivity (giving less weight to the country's acquisition of fertilizer plants).

Agents may be similarly used to generate alerts. News that a pilot has been shot down, for example, may trigger a requirement for local intelligence, spur greater liaison with friendly forces behind the lines, allocate more channels for the pilot's signals, prepare for undertaking search and rescue as well as medical evacuation, and establish stay-away zones for certain operations. Some second-order effects may need to follow automatically (e.g., telling rescue teams which substances the pilot is allergic to).

Engineering a system for agents raises difficult questions. How are they authenticated, made safe for circulation, and kept free from contamination? What access to databases, processor time, and bandwidth are they granted? How explicitly should their logic be documented? Under what circumstances can they trigger other processes, and how can their influence be overridden?

Knowledge management is the search for ways to transfer knowledge. Traditionally, analysis determined the one right way to do things, and that method was then fed to everyone. Although this made sure everyone worked to standards, it stifled initiative, demoted the acquisition of tacit knowledge, and left organizations inflexible to change from the bottom, acquired, in part, through frontline contact with new realities. A contrasting approach would be to liberate everyone to learn the best way to do things for him/herself, but this would lengthen learning curves, retard the diffusion of new knowledge, and lead to problems when skilled workers and engineers retire.

Enterprise architectures are also being asked to support knowledge management—i.e., the art of circulating know-how. Knowledge management was a key factor, for instance, in the U.S. Navy's ability to chase U-boats from the East Coast in 1942.[8] Each commander's

---

[8]See Eliot A. Cohen and John Gooch, *Military Misfortunes,* especially "A Failure to Learn: American Antisubmarine Warfare in 1942," Free Press, New York, 1990, pp. 59–94.

contacts were collected and used to build a database of enemy tactics and successful countertactics. What mechanisms are there to tell those with specific knowledge of problems about acquired knowledge? Writers on knowledge management emphasize the human factors (e.g., trust, random interactions) and the sense that the transfer of tacit knowledge requires physical contact.[9]

Can militarily relevant information be systematically organized? After all, its domain is limited (compared to knowledge as a whole) and routed through well-understood task structures. Much information is generated solely to do specific jobs and can be inculcated through training. Those who repair C-17 engines, for instance, can be handed all the information they need. Nevertheless, modern militaries are continually hiving off more specialized offices, few of which are sufficiently well defined to be exclusively self-contained. Even well-established entities generate new information sources as they wander afield from their original tasking. Also, not every useful data point is contained within the manual; the Web has revealed the power of horizontal communications as a device for getting advice, perspective, and solutions to unique facets of problems. If and as defense systems resemble their commercial counterparts—a clear trend in command, control, communications, computers, intelligence, surveillance, and reconnaissance (C4ISR) equipment but less so elsewhere—a growing share of knowledge on how to manage them will be outside DoD. Finally, open-source intelligence is growing as a percentage of all intelligence. Warfighting is an outside activity often in and among third parties. The ability to ferret information from the environment may come to depend on foraging through a thickening forest of digital data. The ability to harvest such information—regardless of its mixed quality—may provide insights denied to those who abjure such disorganized chatter.

To circulate information intelligently, one needs a way to find where it resides, which gets harder as complexity complicates the task of pigeonholing information into predefined categories. If information is organized in many ways (and much information can only be categorized and catalogued approximately), the possession of a reli-

---

[9]See, for example, Thomas Davenport, *Working Knowledge*, Harvard Business School Press, Boston, MA, 1998; and Nancy Dixon, *Common Knowledge*, Harvard Business School Press, Boston, MA, 2000.

able discovery mechanism is important. Although there are many philosophies for building and maintaining such engines for finding random information (especially machines that can be taught to extract meaning from text), the explicit tagging of a document's subject and key data appears to be growing popular.

Security

Information systems have security architectures that govern

- Who can read from or write to which files (or invoke which processes);

- How legitimate users are authenticated;

- How files and processes are protected from mischief;

- What procedures are used to detect when a system is under attack, respond to limit the damage, and recover functionality when control is restored.

Security's first question is, as always, How much? Systems with insufficient security necessarily depend on the kindness of strangers. But excess security costs money, hassles users, often denies service to the legitimate, and is prone to failure if users react by subverting its rules. Systems that deal with the public (e.g., e-commerce sites) must begin by assuming all users to be legitimate, only rejecting such claims after seeing bad behavior. This rule makes them heir to a flood of messages that are individually benign but collectively choking (e.g., the February 2000 distributed denial-of-service attack). Because DoD systems rarely need open themselves to others, they, presumably, can be locked tight. But collecting information from third parties, winning their cooperation, and coordinating everyone's activities with theirs may sometimes require that at least some military systems become transparent and accessible to those third parties. Barriers will then be needed between those systems and the more closed components of the GIG.

At the second level of protection, the question is one of compartmentation. What topics should be walled off or compartmented? What is the cost of inhibiting the cross-fertilization of ideas? Who decides these questions? Convenience and custom let data owners choose, but is that always best? Should certain types of information have a shelf life, and, if so, by what means can the shelf life be imposed without impeding the interoperability of fleeting data with more conventionally permanent information?

Coerced insecurity creates a third set of questions. What if field operatives with a live tap into the GIG find themselves on the wrong end of an AK-47? Soldiers may be instructed to disable their units prior to capture, but they may not always be able to. Adversaries, unimpressed by the Geneva Convention, can credibly threaten people who shut down their access after capture. Do architects therefore limit what such terminals can access? If so, it is then difficult to see how to develop new warfighting techniques, such as swarming, that require warfighters to be fully and minutely informed about the location, activities, and plans of their colleagues—a prerequisite to tight and time-sensitive coordination of fires and maneuver. Another approach may be to make the GIG's servers sensitive to and thus unwilling to answer people who display the biological indicators of stress that being captured might induce. But might not the stress of combat alone then induce a cutoff of information? And what about giving users alternative passwords (as has been mooted for ATMs)? Will they remember to use them without hesitation? What kind of alternative information can be served up that is realistic enough not to anger the captors and put the captured at risk but that also does not reveal compromising details? How can adversaries be shown deceptive data without our own forces being fooled?

Trust is the fourth question. Should people be vouched for centrally or through intermediaries who presumably know for whom they vouch? Should access to the system by its components be predicated on the diligence with which they defend computers? If so, how can such diligence be effectively measured? Should architecture be transparent so that users can understand it, or opaque so that foes cannot?

Interoperability

Interoperability is needed on several levels. At the bottom is the ability to pass bits physically. Next is syntax, the ability to format bits into discrete manipulable chunks (e.g., TCP/IP, HMTL, JPEG). Its faults notwithstanding, DoD's Joint Technical Architecture has pointed the way for interoperability at the physical and syntactic levels. The invention and widespread acceptance of XML as the markup language has resolved the syntax problem of and thus set the stage for the third level: domain semantic standards—the ability to refer to the things and concepts of the real world in mutually understood ways. DoD has endorsed a clutch of semantic standards for e-commerce, message traffic, and mapping symbology, but many domains have no standards, and others have far too many that conflict with each other. Numerous databases use similar terms in incompatible ways, a problem informed but hardly solved by the ongoing conversion from database schemas to XML schemas. The fourth level deals with process semantics: standard ways to express intentions, plans, options, and preferences so as to support negotiations.

The quest for semantic interoperability has architectural ramifications. Behind every semantic structure is an implicit data model of the world, one that makes some things easier to express and other things harder. How the structures of *human* language influence thought is a long-standing issue in linguistics, with some seeing great influence.[10] Whether similar semantic structures exist for digital data is even less clear. Computers are less flexible than people and are thus more in thrall to their limited vocabulary. But to what extent will people take their cognitive cues from their machines' semantics?

The standards *process* within an organization also reflects the organization's distribution of power. Standards can help consolidate the current GIG, which is composed of stovepipe systems managed by disparate owners. Open systems can also help DoD share information with coalition partners and their systems. Can stocking the GIG with interesting information and useful applications that use standard semantics induce others to pick up such standards for their own interactions? Or, alternatively, will various semantic constructs bubble up, with interoperability generated through explicit negotiation over what terms mean or the implicit use of inference?[11] Should

---

[10] An example is the appendix to George Orwell's *1984*. Also see Steven Pinker, *The Language Instinct*, Harper-Collins, New York, 1995; or Noam Chomsky, *Aspects of the Theory of Syntax*, MIT Press, Cambridge, MA, 1965.

[11] See Tim Berners-Lee, *Weaving the Web*, especially Chapter 13, "Machines and the Web," Harper-Collins, New York, 1999.

there be a common framework for referring to real-world objects, or should there be multiple frameworks, each called out by name?

Interoperability issues arise in great force when coalitions are being formed with partners that can build their own global information systems. How should the architecture of something akin to a NATO integrated system be developed? There are no easy choices. Having the United States and Europe build separate systems and merge them as they mature ensures that end-of-cycle integration will be long and hard. The United States could supply the global components (such as long-lines, satellites, and long-endurance UAVs) and make these the background against which local (and thus often coalition-supplied) components sit. Or the United States could supply an architecture in sufficient detail to make European systems plug-and-play. Both are technically feasible approaches, but the Europeans are unlikely to be happy playing second fiddle. Having the United States and Europe build a NATO architecture together from scratch could delay the U.S. architecture by five to 10 years. Standard, or at least explicit, interfaces in planning the U.S. GIG cannot hurt—regardless of which of these four choices is made.

Can access to DoD's GIG be used to induce third world nations to become allies? Although technical issues may be less daunting in linking systems in this case—DoD will give, third world allies will take—trust issues are more salient. Today's friends may not be friends tomorrow, so the case for giving them deep access is hard to make. What are the terms of the relationship? What will such friends be shown, with what restrictions, and in exchange for what? Should the United States applaud or even seek to induce changes in these partners' information systems (e.g., their forgoing the purchase of potentially competitive C4ISR systems or their adopting an architecture and thus viewpoint similar to DoD's)?

Indeed, how deeply should one nation's military information system penetrate another's? Take Walmart's relationship to Proctor and Gamble, one of its major suppliers. Walmart could have chosen to collect data on its own sales, in-store, and warehouse inventories and then calculate the restocking schedules and hand the resulting orders over to P&G. Instead, it makes intermediate data directly available to P&G and leaves P&G responsible for scheduling production so as to optimize Walmart's inventory of P&G products. In deep partnering relationships, all partners can write to the files of the others—much as several doctors may enter information on a patient who is under the primary care of a specific physician—and can trust imported information enough to automatically fuse it with their own. Partners may contribute to a library of codes and subroutines to be liberally passed around. A further step could be for partners to populate a jointly accessed architecture with applets, servlets, agents, or knowledge rules, each capable of successively transforming raw information. When systems seek mutual intimacy, their success may depend on having a transparent architecture.[12] But all these are explicit policy choices.

Integration

Systems must work together. But designing an architecture to make them functional on day one at the expense of everything else is a recipe for growing obsolescence on day two and beyond. How and when should capabilities be added? Is it better to have everything up on day one, or is it better to introduce capabilities (or at least major components thereof) serially, shaking out the bugs one stage at a time? The second of these, incrementalism, has two problems. First, a tool such as a grammar checker must reach some threshold of capability before it is worth using at all. Switching has costs, and users forced to switch too often will simply balk. Second, incrementalism can lead to a hodgepodge. Legacy features (such as DOS and the four-byte address space of IP, not to mention two-digit years) may be hard to eradicate. Conversely, projects that aim to have everything up all at once force users to wait a long time. Only a third of all complex systems arrive on time, and failure can be devastating. And even success can be accompanied by great weariness and hence a wariness of change, or can mean a structure so finely tuned to today's problems that it resists conversion to tomorrow's. All-at-once complex systems are increasingly ill adapted to a world that is always moving on.

---

[12]For example: to provide targeting guidance to a topside gun on an Italian frigate, data from a British UAV's electro-optical sensor is linked through a U.S. network to readings from Dutch microphones, the data flows being fused with the help of a French-hosted software agent and being compared to a German-provided database of marine templates.

Another key issue is whether users feel the systems are theirs, in the sense of being a tool they understand how to use and apply to their problems. Tools, as often remarked, should feel like a natural extension of the body or mind. Good carpenters or musicians take proprietary and personal interest in the tools they use, up to the point, sometimes, of not allowing others to touch them. Tools that are balky, arbitrary, fussy, and unreliable are less likely to be used and, if forced on people, less likely to be mastered or maintained. Will the GIG be perceived by users as friendly and trustworthy enough to entrust their lives and time to? Will they come to "own" their piece of it, seeing it as something they have control over and therefore will step up to feeding and caring for?

THE NEED TO THINK NOW

Information systems reflect the organizations that buy, run, and embed them in an architecture. The architectures reified in computer and communications devices tend to replicate, with some mix of conscious design and unexamined assumptions, relationships that exist among people. But architectures also have a funny way of molding these relationships. Information systems are a tool of amplification and disenfranchisement and inevitably alter the balance of power in any organization they enter.

In their dawning, computers were strange creatures, each seemingly programmed with its own language and conventions. As computers became standardized in the 1960s around the IBM 360 mainframe, the popular expectation was that they would become instruments of top-down control. Data would be entered at the bottom, collated, organized, drawn upwards, and used to give top-level managers a much finer picture of their enterprise. The autonomy of middle managers, which was based on their possessing knowledge too variegated to be effectively amalgamated, would thus end, as would the workers' freedom of action, which was based on their being able to make their own decisions. Computerization into the 1980s did, indeed, encourage the amalgamation of disparate enterprises into conglomerates managed "by the numbers" alone.

However, when the proliferation of personal computers in the workplace that began some 15 years ago combined with the privatization and rapid spread of the Internet that began in 1992, a revolution from the bottom was sparked. Once again, there is an expansion of approaches and a sense of great change, with considerable disagreement about what these all mean for human contact. Will this prove permanent or temporary? Again, the disenfranchisement of middle management appears imminent. In all this flux, it seems right to empower everyone to search for the new best way or at least the best way for him or her.

It is too early to tell whether this empowerment represents a revolution from which there is no going back or simply the blooming of a thousand flowers before a new reconsolidation. But it is not too early to think clearly about the situation. DoD has had, does have, and will continue to have an information architecture, regardless of whether it knows or admits as much. This architecture is driven in large measure by policy choices; it is not solely a consequence of information technology. These policy choices should be made explicit and maybe open in most cases. Otherwise, DoD may either unconsciously make choices it does not like or subconsciously opt for immediate efficiency over longer-term adaptability.

PART II. COPING WITH UNCERTAINTY

INTRODUCTION TO PART II

Defense planning involves a host of factors that interact with each other over a time period often measured in decades. The first M-1 tanks and the first F-16 aircraft entered the force more than 20 years ago, and most of the Navy's capital ships stay in the force for 30 years or more. The longer the time horizon, the harder it is to know the parameters of a decision with any precision. At any point, there are "knowns"—things people know they know; "known unknowns"— things people know they do not know; and "unknown unknowns"— things people do not know they do not know. The deeper the reach into the future, the more the unknown unknowns dominate.

Like all humans, defense planners exercise what is called "bounded rationality."[1] In other words, they lack complete knowledge and anticipation of the consequences of their choices and can think through only a few alternative courses of action on their own. They cannot anticipate future consequences without actively using their imagination, and the imagination of any individual is limited. Faced with complexity and uncertainty, individual planners risk becoming comfortable with familiar mind-sets or illusions. Group decision-making, in turn, risks producing "safe" decisions as members march down the path of "groupthink" to shore up their positions.

No one denies that uncertainty is important and that planners should deal with it as best they can, but the full extent of the problem of uncertainty is not often appreciated—even by planners themselves. Like most of us, planners seldom go back to compare as-

---

[1]A term associated most notably with Nobel-prizewinning economist Herbert Simon.

sumptions they made years ago with what actually happened. Instead, they just go on, vaguely aware of having adapted to circumstances but not at all aware of the extent of their adaptation. Large organizations may be even less aware of adaptations that have proven necessary and even less humble when laying plans for the future. This failure to look uncertainty in the face is perhaps less evident now because of the shock September 11's events brought to U.S. foreign policy and defense planning. Everyone is well aware of uncertainty at this point. In the longer run, however, the tendency to sweep it under the rug will reappear. It is, after all, a natural human tendency.

Moreover, rarely will any one defense planner or decisionmaker possess all the kinds of knowledge and experience needed to face uncertainty and still make good choices. The quest for good decisions thus drives planners and decisionmakers to find tools that can help them cope with the many conditions of uncertainty.[2] Fortunately, there are techniques that can help test the robustness of the knowns, put some bounds on the known unknowns, and discover and even illuminate the unknown unknowns.

The first chapter in Part II, Martin Libicki's "Incorporating Information Technology in Defense Planning," starts with the premise that while the future is not knowable, neither is it a complete mystery; a few educated guesses about the future can go a long way to help planners. Even though many decisions, such as those about force levels, can be reversed in the short term, the future matters precisely because of the long shadows cast by decisions about weaponry or research and development (R&D) or precedents that might be set by policy. One good starting point for the analysis is broad trends that *are* discernible, such as demographics and the continuous improvement in information technology. Libicki works through what this trend in technology means for future conventional combat and information warfare. Improvements in information technology, for instance, suggest that high-intensity conventional warfare could entail

---

[2]For more detail, see Aaron Wildavsky, "The Self-Evaluating Organization," *Public Administration Review*, No. 32, September/October, 1972, pp. 295–365. Also see Paul R. Kleindorfer, Howard C. Kunreuther, and Paul J.H. Schoemaker, *Decision Sciences: An Integrative Perspective*, especially Chapter Six, "Group Decision Making," Cambridge University Press, New York, 1993.

feeding copious amounts of sensor-derived information to a common picture, which, in turn, would be used to determine what to shoot at. Moreover, increasing dependence on information technology does not necessarily make information warfare more attractive for an adversary. Indeed, the very multiplication of information sources makes users less vulnerable to attacks on any one source.

Chapter Five, "Uncertainty-Sensitive Planning," by Paul Davis, begins with a method first developed a decade ago for going well beyond conventional wisdom in contemplating the future. It identifies not only a "no-surprises future," but also various types of possible uncertainties, and then develops plans to cover some contingencies explicitly and to hedge against others. The method also emphasizes environment shaping, based on the argument that the future can be shaped to some degree by U.S. actions. Davis combines this idea with the handling of multiple objectives to define a portfolio-management framework for high-level defense planning. He then applies this framework to the problem of developing capabilities for a wide range of political-military scenarios and operational circumstances.

The post–Cold War focus on one or two illustrative planning scenarios for defense decisionmaking is no longer consistent with today's goals of flexibility, adaptiveness, and robustness. Moving toward such goals requires that alternative force postures be evaluated in an explicit "scenario-space framework" wherein their value can be measured by the variety of circumstances in which they would be effective. This approach to capabilities-based planning is sharpened by the discipline of working within a budget. It thus forces choice. The last part of the chapter describes an analytic framework for combining the portfolio-management construct of overall defense planning with the results of capabilities-based analysis and economics.

In Chapter Six, "Planning the Future Military Workforce," Harry Thie addresses manpower, personnel, and training issues. How are personnel managed—directly, through assignments, or indirectly, through incentives (such as compensation)? How are skills to be transmitted and behaviors to be inculcated? How large a force should be raised? What should its grade, skill, and experience composition be? What are the best ways to procure, enter, train, develop, assign, advance, compensate, and remove people? The key manpower issues have shifted with America's changing security circumstances. As America entered World War II, the issue was how to procure a large force immediately. During the Cold War, it was how to manage a large inventory of people with military experience. External events, societal concerns, missions, organization, technology, budget, and demographics, in turn, shape particular subissues: recruiting, training, retaining, promoting, compensating, and retiring. Against this background, Thie inquires about the personnel and training policies that might best achieve a force that is big enough, qualified, stable, experienced, and motivated. Some policy choices can be applied servicewide; others apply to military units or to individuals. The chapter illustrates how both controllable and uncontrollable variables give rise to complexity and conflicts among competing objectives, as well as to potentially unwanted outcomes.

In Chapter Seven, James Hosek describes how during the past 30 years, the experience of an all-volunteer force has led to a remarkable accumulation of knowledge about the importance of personnel quality for military capability, and about the policies for getting, keeping, and managing such a force—what Hosek calls "the soldier of the 21st century." The U.S. military has been at the forefront in recognizing the value of human capital, and RAND has systematically explored alternative policies for efficiently managing human resources. In the future, the value of people and of the knowledge and skills they possess will become even greater. The story of the 21st century soldier is thus not only about understanding why high-quality personnel are vital to the U.S. defense capability and what can be learned from the history of the volunteer force, but also about what considerations are key in the seemingly mundane, yet fundamentally crucial, task of setting personnel management and compensation policies.

The last chapter here is Frank Camm's "Adapting Best Commercial Practices for Defense." Best commercial practices are those that lead to better, faster, and cheaper products in the companies where they exist. DoD can turn to these practices in determining requirements, designing processes, selecting and making use of external sources, and managing ongoing performance. But to identify those practices with potential for DoD use, DoD personnel must systematically compare the department's performance with that of exemplar organizations of all kinds. As DoD thinks about adapting a specific practice, it must remember that the differences between the commercial setting where the practice was observed and the DoD setting where the practice may be used are crucial. Differences in organizational culture, priorities of stakeholders, and the structure of major management systems affect outcomes. Once DoD understands these differences, it can examine the barriers they present and adapt the practice to overcome them for the DoD setting. DoD should then approach the adaptation of a best commercial practice as a variation on organizational change, addressing all issues likely to arise when any new practice enters an organization. From this perspective, DoD will find that *how* it pursues a best commercial practice is often more important to success than *which* practice it pursues.

Chapter Four

INCORPORATING INFORMATION TECHNOLOGY IN
DEFENSE PLANNING

Martin C. Libicki

Inherent in the human condition is the fact that although we will live the rest of our lives in the future, every decision we make is based on what we have learned from the past. For defense planners, this is more than a nominal or philosophical conundrum. Those who plan defense programs face the very real possibility that the world in which these programs reach fruition will be different from the one in which they were planned.

This chapter introduces some guidelines for, if not predicting the future correctly, then at least coming closer to a correct prediction than do those who unconsciously assume the future to be equal to the present. We can think logically, but only time will tell how correctly, about the future. This fact is demonstrated by taking something that can be predicted—the information revolution—and thinking through its likely effect on conventional warfare and the extent to which new forms of warfare, such as information warfare, make sense.

MODEST PROPOSITIONS ABOUT THE FUTURE

Those who fear error should avoid forecasting. Many an expert has been famously wrong about prospects of one or another technology. Yet defense planners are called on to make decisions that will depend on the world's state 10, 20, and 30 years out. Undertakings that bear fruit years hence—long-lived investments, promises hard to back out of, self-reinforcing institutions, standards, and visions—are examples of long-range decisions that have yet-to-be-determined outcomes.

Forecasters make mistakes, it is true. But the effort of forecasting is still valuable if it makes *explicit* statements about the future that are closer to reality than the *implicit* assumptions that too often guide long-term policies. But how is this to be done? Naive forecasters are heir to three clichés: that change is linear, that the pace of change is accelerating, and that complexity is thus increasing.

Not all trends run forever. The rapid progress that characterized aerospace, from Kitty Hawk to the moon landing, was universally expected to continue; it was commonplace for equipment deemed revolutionary upon its introduction to be scorned as obsolete 10 years later. Yet this was not what happened. People did not have an exceptional need for great speed, and by the late 1960s there were few radical technologies that had not yet been discovered. So progress slowed down. The SR-71, the F-15, the Boeing 747, the Concorde, and the Saturn may have been surpassed, but not by very much even *30 years later*.[1] Yet *technical* progress made through 1970 still echoes in terms of *social* and *economic* changes. Pacific air travel, for example, still doubles every seven years because cheaper, faster travel has fostered global institutions and linkages that, in turn, drive the demand for more travel.

Information technology has now become hot, but it is easy to forget what has cooled in the meantime. Today's office environment would be unrecognizable to someone arriving fresh from 1979, but many of the computer systems that most affect our lives—for business management, financial transactions, process control, and transportation scheduling—have been around for twice as long. Indeed, in many ways, overall change may be decelerating. Boomers born in the 1950s and 1960s had a different childhood than their kids are having, but the differences between boomers and their parents are greater—whether measured by income, education, transportation, mass media, or the likelihood of reaching age 60. And the social impact of the

---

[1]The same can be said for plastics (see *The Graduate*, 1968). Every bulk plastic in use today had already been invented by then. Within 10 years, growth rates in demand had fallen from 10 percent to 3–4 percent—just slightly faster than the overall economy.

birth control pill, introduced in 1963, has yet to be exceeded by anything in biotech.

The reason the past often appears simpler than the present is that great issues, once resolved, lose their complexities. In some ways, life is simpler now. Today's personal computers are easier to use than those of the mid-1980s, which, in turn, were far simpler than those of the mid-1970s. Automatic transmissions are simpler than stick shifts. Airline schedules are easier to understand than train schedules.

Having shed the clichés of naive forecasters, the next step is to wring as much as possible from domains where the future is clear, only then moving to the speculative. A fruitful path from easy to hard starts with demography and works through technology, the environment, and economics before chancing the social and political realms. Such a flow follows causality. Technology changes society, but society can only modestly alter the vector of technology. After all, technology must obey physical law.

Demography is a good place to harvest the unexpected from the inevitable. Clearly, for instance, the number of 30-year-olds in the year 2025 cannot exceed the number of 8-year-olds alive today (2003). Better yet, because most 8-year-olds have a high likelihood of reaching 30, and most will mature in the country where they are born, the number of 30-year-olds in 2025 can be forecast with some confidence. This simple generalization suggests that greater Europe, which had four times the population of the Middle East in 1978, will have fewer 30-year-olds come 2025. Even if the long-standing birth dearth in Europe and Japan ends, the size of the productive age group will shrink unless there are large immigration flows—inconceivable in Japan, and possible only at the expense of great change in Europe.

In *technology*, the most fundamental theme over the past 25 years has been the shift from ever-larger to ever-smaller as the driver of change. Progress used to mean size: world-scale factories to serve global markets, taller buildings, heavier supertankers, wider roads, longer runways, gigantic rockets, and multimegaton warheads. Circa 1975, energy shortages, pollution, integrated circuits, and scanning tunneling microscopes heralded a new direction. Today's new factories, buildings, supertankers, roads, runways, rockets, and warheads are not particularly larger than they were 25 years ago. Progress since then has come from the ability to engineer features and to control defects at the micro scale: microelectronics, microbiology, and micro-electromechanical systems (MEMSs).

Microelectronics—measured via processor speeds, communications throughput, and storage capacity, all now doubling every two years or faster—should continue to make progress. Personal devices will continue to be the prime beneficiary, especially with improvements in untethered energy sources: efficient batteries, miniature fuel cells, and photovoltaic devices. Major appliances may soon be network-ready right out of the box, perhaps even looking to link to any network they can find.

As for microbiology, with the human gene now being read and its mapping into folded proteins to follow, scientists are learning a great deal about how life works, how life fails, and the pathogenic and genetic correlates of disease. Easier and earlier disease detection will yield more effective treatments. Research on the human stem cell suggests that organs for transplant will be grown rather than acquired. Such techniques are close to what human cloning requires, making it a near-certain event.

Microstructures are apt to proliferate and become more useful. MEMSs, whose structural features are similar in scale to prior-generation electronic chips, are useful for detecting movement as well as subtle visual, thermal, acoustic, and biochemical features of the environment. Microwatt transmissions from small devices have been demonstrated. So have very small combustion chambers that yield power devices with 10 times the energy density of batteries.

All of these technologies can be used in sensors that have potential military applications: small cheap microphones, electro-optical charge-coupled devices (CCDs), biochemical detectors, and pocket radars for security, biomedical, and controller applications. Sensors the size of pennies may be littered across the battlefield, coupled with MEMS-aided transmitter-receivers that can operate in a remote area for weeks or months. Biosensors on chips or natural sensors on insects may find suspicious chemicals in the air or on surfaces. This confluence suggests that sensors will be key to what tomorrow brings.

Technological vectors have underlying causes. The ability to scale up physical devices reflected steady improvement in knowledge, which, at some point, reached diminishing returns. Nothing so important, one is tempted to conclude, is likely to be realized from macro-scale technologies. No such diminishing returns are apparent from the trends toward miniaturization. Cheap electronics, for instance, fed a global demand for communications, which spurred the launch of communications satellites, which faced a primary constraint of spectrum, which is being relieved through techniques such as beam forming and phased array antennae, which led to a secondary constraint of power radiation, which is spurring the next generation of 25-kW satellites. If technology favors small things, it will favor many of them, and the problem of controlling multitudes will loom. With only a dim prospect of radical advances in the writing and debugging of software, complexity theory, which posits that the right rules can induce complex behavior, can prove its value for simulation and control.

REAL REVOLUTIONS ON THE PHYSICAL BATTLEFIELD

So are there fundamental changes under way in warfare? And, if so, what are their essential characteristics and limits? In 1978, Under Secretary of Defense William Perry noted that DoD would very soon have the ability to see everything of interest on the battlefield, hit everything that could be seen, and kill everything that could be hit.[2] "Very soon" always has to be taken with a grain of salt, but Perry *was* on to something important.

Hitting What Can Be Seen

What does precision mean? Although a tank within 3 km of its target is a precise weapon, attention has rightfully focused on the precision guided munitions (PGMs), which come in three basic types, distinctions among which show how information technology may influence warfare.

---

[2]From Philip Morrison and Paul F. Walker, "A New Strategy for Military Spending," *Scientific American*, Vol. 239, No. 4, October 1978, pp. 48–61. For another early treatment, see "The New Defense Posture: Missiles, Missiles, Missiles," *Business Week*, August 11, 1980, pp. 76–81.

Man-guided PGMs require that the targeter be within visual range and thus a few kilometers of the target. Examples include the TOW (tube-launched, optically tracked, wire-guided) antitank missile, and laser-guided artillery shells and bombs (which date from the end of the Vietnam War). Although the reliability of the PGMs is usually good, their accuracy is only as good as the targeter (who can be distracted by being shot at).

Seeker-guided PGMs find targets using their own sensors. Early types included radar-guided anti-aircraft missiles, heat-seeking missiles, and acoustic torpedoes. The United States has PGMs for acquiring and tracking many types of signatures, and these weapons can usually be fired from standoff ranges. But only so much sensor capability can be put in a small package, and miniaturization that exceeds what is commercial practice tends to be costly. As a result, PGMs are expensive and thus, under current doctrine, available only for use against high-value targets.

Point-guided PGMs, directed to a specific location in real space, include ballistic missiles that rely on inertial guidance, terrain-following cruise missiles, and, more recently, PGMs guided to a specific latitude and longitude by reference to global positioning system (GPS) signals (the Tomahawk [Block IV] cruise missile and the Joint Direct Attack Munition [JDAM]). DoD supposedly lacks munitions that go to a moving point on a map, but there is no technical reason why they cannot be produced. It helps that the United States is working hard to map every object to within 5 or 10 m in absolute coordinates and 1 m in relative coordinates. GPS satellites make it possible for a receiver to know its location to within 18 m, and far closer relative to reference points. If the locations of the target and the PGM seeker are known, the seeker's course can be programmed so that it hits the target.

Why belabor these distinctions? Because they have everything to do with where the "smarts" must reside within the system. Does the sensor go in with the targeter, inside the weapon, or remain external? Giving PGMs their own sensors makes them more robust and shortens the sensor-to-weapon loop. But with limited sensor capability, thanks to cost and carriage factors, they are more vulnerable to deception and have difficulties acquiring targets at long range. The use of off-board sensors to generate target tracks not only provides improved range, tracking, and identification, but also complicates strategies for PGM deception. But point-guided PGMs need a reliable link to both sensor information and the overall command-and-control (C2) system.

Ironically, this technical disadvantage—the need for a reliable link—has political advantages. For instance, the United States supplied surface-to-air missiles (SAMs) to the Afghan rebels in the 1980s and, by doing so, lost control over them. When the Soviets left, the United States started worrying that the SAMs would appear as terrorist weapons.[3] Had these weapons needed externally generated flight track information to work, the United States could have removed this information to devalue them. With self-contained PGMs, power rests with the operator; with point-guided PGMs, it rests with those who control the intelligence system. PGMs, notably point-guided PGMs, also shift the locus of power internally. It is a simple fact of war that if one can kill everything one can see, it pays to see as much as possible of the other side and, equally important, to keep oneself hidden. This ability to seek and to hide, more than the ability to mobilize forces and deploy them into battle, determines outcomes, and thus determines which job contributes more to military outcomes. Here is an application of exponentiation. The existence of precision weapons creates a need for precision information.

The process of seeing is transformed by the growing profusion of sensors in space, on aircraft both manned and unmanned, on and under water, and on the ground. The first effect of using many sensors rather than relying on one super-sensor is the potential for great flexibility in deployment. A penetrating aircraft can survive if it flies below radar, but if microphones are placed under the aircraft's probable path, defenders can detect where the aircraft was flying and perhaps even track it. Flexibility permits one to adapt to the operational challenge of conflict. If the nature of conflict and of the enemy changes, sensors ought to change as well. The variety of sensors becoming available makes it more likely that the battlefield can be made so transparent that the first part of Bill Perry's vision can be realized. Yet transparency will vary greatly. On water and in the

---

[3]The SAM aimed at a commercial airplane flight out of Kenya (November 2002) was actually a Russian model.

desert, nothing large can be hidden for long; on plains and chaparral, some cover is available. Forest and jungle, and finally cities, are progressively more opaque. But in cities, the searching, while more difficult in some aspects thanks to greater cover and the presence of civilians, can be partly accomplished by friendly residents who supply eyes to do much of the work.

The second effect of quantity is that it has its own quality. DoD has many sensors that can illuminate the battlefield with a high degree of precision. Satellites in low earth orbit can take very accurate photographs, conduct strategic reconnaissance, and chart static formations, but they cannot track and target moving objects. If acuity requirements can be forsworn, the purchase and use of many satellites would permit continual, perhaps continuous, target tracking from space. And unmanned aerial vehicles (UAVs) can provide both accuracy and continuity. The Clementine spacecraft, launched by the U.S. Naval Research Laboratory (NRL), proves that many good sensors can be placed on a $50-million satellite. In theory, hundreds of cheap satellites could, along with UAVs, provide continuous coverage for spotting and tracking fleeting targets, at least under favorable weather and lighting conditions.

Another effect of using many small sensors (and one that may ultimately prove to be the most decisive factor) is that it offers better survivability. Having many small sensors rather than a few large ones lessens sensor vulnerability as would-be foes become able to see the battlespace better—an example of how two-sided change carries far different implications than does one-sided change. JSTARS (Joint Surveillance and Target Attack Radar System) and AWACS (Airborne Warning and Control System) are extremely capable sensors, easily the best in their class. But both are mounted on large and not terribly stealthy aircraft. They survive by flying behind front lines, which works only until enemy missiles can range hundreds of kilometers, or enemy shooters can penetrate front lines in the air or on the ground.

A constellation of UAVs would be more able to survive. An individual UAV is not as capable as either of the two aircraft, but UAVs could collectively illuminate the skies. UAV suites include the electro-optical, infrared, synthetic aperture radar, passive millimeter wave, and light detection and ranging sensors. Similarly, although an Aegis cruiser can see surface targets tens of kilometers away and air targets perhaps ten times farther, it is, compared to aircraft, even less stealthy, more complex, and costlier, and it may some day compete with buoy-hosted sensors that individually are less capable but collectively could be less vulnerable. Ground sensors—microphones, remote cameras, and other sensors that can measure seismic, gravitational, biochemical, or magnetic phenomena—may be the ultimate in distributed searching. Some of the detectors coming out of the medical field are small, precise, and intelligent. Ground sensors in development are expensive, but commercial technology suggests that far lower costs are feasible. The wholesale cost of a PC-mounted camera is, for example, about $10; the wholesale cost of a microphone is well below that of its batteries.

The fourth effect of using many sensors is that it puts a premium on data coordination, correlation, and fusion. One sensor may excel at locating an object but not at identifying it. Another may be able to distinguish whether an object is a pickup truck or a tank but not be able to find the object precisely. Using different sensors not only reduces the uncertainty of where the object is, but also helps identify it more confidently. With the importance of sensor coordination across not only phenomenologies but also media, there is no good alternative to thinking about battlefield illumination in a joint context. Jointness is not simply a matter of warfighters with different-color uniforms working together; it now includes asking their machines— which are a good deal more finicky and far less clever—to do so.

If one can see from afar, why shoot from up close? Targets can be hit from 20 km with air-launched munitions, unattended remote-controlled weapons, or pop-up, or even shoot-and-scoot, platforms that are hard to shoot back at. Targets can be hit from 200 km with weapons such as the Army's Tactical Missile System (ATACM), stealth aircraft, and UCAVs; and they can be hit from 2,000 km with ballistic and cruise missiles. Are PGMs too expensive for most warfare tasks? Today's cruise missile costs $600,000. The United States fought the Vietnam War as a war of attrition (wisely or not), spending $1.2 to $1.5 million to kill each enemy soldier. If, as is typical in aerospace manufacturing, each doubling of quantity lowers unit costs by 20 percent, cruise missiles purchased at the rate at which enemy soldiers were killed in that war would cost $100,000 to $200,000 each. Using even expensive weapons to kill three people around a campfire hardly adds to war's costs. Reaching farther than the enemy can reach back is a comparative advantage even—especially—in a perfectly transparent world. And the ability to build long-range propulsion systems is likely to be something the United States and its allies can do well and likely adversaries cannot.

The need for standoff range illustrates today's strategic asymmetry between the United States and its likely opponents. In a slugfest where dollar is traded for dollar, the United States will come out ahead—it can expect to enjoy ten, and more typically, one hundred times more gross national product (GNP) than any of its likely opponents for a long time to come. But if the United States finds itself trading life for life, it will lose.

From Contingency to Necessity

What is suggested when one puts sensors and weapons together is that modern militaries should wage war by using sensors to find targets and using PGMs to prosecute them. But *should* does not imply *will*. The United States and its allies can also go to war in the old-fashioned way—mobilizing force against force—and still do fairly well, as Desert Storm illustrated. But for how long? Which is to say, how great a divergence can arise between what technology promises and what its users grasp?

In Desert Storm, the coalition went to war against Saddam Hussein, an enemy who was fairly blind to what information technology could do. Previous RMAs (such as the dreadnought, blitzkrieg, and nuclear weapons) were built on items not found at local stores, but the current RMA is based on what comes from commercial sources as well as what comes from military labs. What if an adversary opened a checkbook and bought PGMs from the French or the Russians; UAVs made by any of 20 countries, plus digital cameras or digital videocameras (particularly after the successful advent of high-definition television with its thousand-line resolution) to put on them; space-based imagery, terrain data, and software to meld the two into fly-through quality virtual reality sets; digital video disks (today's 4.7 gigabyte drives can hold an image of all of Yugoslavia accurate to 2 m); portable personal computers; palmtop cellular phones; access to global communications; GPS receivers-on-a-chip; and night-vision goggles—all to provide a fine ground-level complement?

Now take Desert Storm and run it against this enemy, which, armed with these information technology tools, is far more sophisticated than Iraq was. The coalition did three big things to win in Desert Storm: ship in a mountain of material, take out Iraq's C2 capability, and run free over the battlefield, first in bombers and then in armor. The same war fought the same way against a sophisticated adversary—maybe not rich but smart enough to wire itself up in advance—may work poorly, for the following reasons.

First, having to ship in a mountain of material is a recipe for disaster, because most of the elements of the logistics infrastructure—ports, bridges, ships, airlifters, and logistics piles—are highly vulnerable and poorly hidden. Logistics can be attacked by volleys of PGMs.

Second, the coalition could cut Iraq's ability to talk because Iraq used centralized systems such as mainframe computers and central office switches. The world today is moving toward cellular phones and local area networks (LANs), a more distributed architecture and one far harder to knock out.

Third, although maneuvering is more attractive than sitting and dying, maneuver entails moving, and moving disturbs an environment, creating signature and thus making one a target for destruction. Walking into Desert Storm II against a sophisticated enemy could mean a great deal of trouble—and this is without even considering weapons of mass destruction (WMD) being used against massed forces.

This raises the question of platforms—a question that is an unintended but logical result of the information revolution. Today, platforms rule: the Air Force is built around aircraft; the Navy, around ships; and four of the five Army combat branches (artillery, tanks, air cavalry, and air defense), around platforms. Platforms have grown more capable, but only by also growing more complex and costly. As they become more costly, fewer are bought; with fewer in the inventory, owners want them better protected. Self-defense systems are thus needed, raising costs further. As the number of acquisition programs under way declines, each community in DoD becomes more desperate to get its requirements adopted in the new platforms— which retards the acquisition cycle, meaning that fewer can be started at any one time—hence, two vicious circles. Norm Augustine, before becoming CEO of Lockheed-Martin, observed that current trends would leave the U.S. military with one aircraft by the middle of the next century: an aircraft flown three days by the Navy, three days by the Air Force, and on Sundays by the Marines.[4] By contrast, UAVs can be cheaper for owners to replace than for adversaries to shoot down.

Distributed sensors and weapons that are redundant and overlapping are not worth risking lives for, because they are built on civilian specifications and thus benefit from great economies of scale. Although it is harder to deal with a mosaic of smaller pictures than with one big one, only computer power stands in the way of converting the one into the other, and computer power is getting cheaper every day.

The Coming Architecture of Military Organization

Information technology suggests that it may be logical to convert from a platform-based to a knowledge-based military (see Figure 4.1). In the former, operators, local sensors, weapons, and self-protection are all bundled as platforms that work together but are each essentially an autonomous fighting unit. They are given intelligence, but once in combat they usually work with data they collect themselves.

In a knowledge-based military, fused data from networked sensors go into creating a shared knowledge base of the battlefield. All of the battlefield appears in low resolution, and some parts of it appear in high resolution; the two views come in and out of focus as need dictates. Such knowledge, in turn, feeds the weapons, because it generates the targets and tracks they are aimed against.

This architecture permits the separation of information and operations, so those who find the target need not be those with the trigger. Such a separation offers opportunities to build vertical coalitions, in which the United States supplies to local allies information they can use to conduct a war. Why build such coalitions? First of all, they

---

[4]Norman R. Augustine, *Augustine's Laws and Major System Development Programs* (rev. and enl. ed.), American Institute of Aeronautics and Astronautics, New York, 1983.

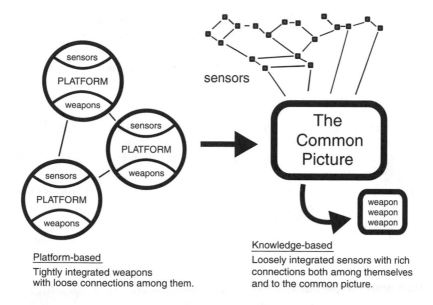

Figure 4.1—The Move from Platform-Based to Knowledge-Based Warfare

would enable the United States to avoid some of the many disadvantages associated with its going to war. For example, the large logistics infrastructures its force structure requires are, as noted, vulnerable, and the large fingerprints left by U.S. operations often work against its political interests. Moreover, if the United States can outfit its friends with PGMs (some have them already), and then provide the requisite information on targets and other enemy dispositions, the friends could prevail in what otherwise would be an evenly matched contest. As an illustration of how this would work, replace "weapons" (lower right corner) with "friends" in Figure 4.1. The United States gives its friends the common picture.

Vertical coalitions also offer other opportunities for altering the nature of the U.S. role and presence. During the Bosnian conflict, NATO was eager to bring Serbians to the bargaining table, but the Serbians would not quit as long as they believed they were militarily superior to the Bosnian and Croatian forces. By presenting the latter with a usable picture of the battlefield, NATO could have tipped the balance against Serbia without setting foot in the Balkans.

In the 1970s, as Egyptian and Israeli forces disengaged, the Sinai was wired to give both sides early warning of an invasion. In the future, DoD could fashion technology to generate not only early warning, but targeting information as well, creating an effective no-man's-land between the two combatants. Violators would show up as targets, putting them at immediate risk.

Information can be the glue behind security arrangements. Countries joining NATO could get a jump-start on interoperability if their C4ISR (command, control, communications, computers, intelligence, surveillance, and reconnaissance) infrastructures and NATO's were compatible beforehand. Were, for instance, Poland and its border areas well illuminated, Poles could feel better about their security without necessarily having to arm themselves heavily or see large numbers of troops deployed on their soil. Information can reassure friends of the United States who eye each other with suspicion. Were Asia better illuminated, each nation could justifiably feel confident that it could see something coming or conclude that nothing was. Belligerence could be further dampened if access to such illumination depended on how each nation behaved and thus postured its military.

The ability to flood the battlefield in lights also carries great implications for force structure. DoD has traditionally scaled its forces for warfare according to enemy strength. But what if the real question is what it takes to illuminate the battlefield, and then how many weapons it takes to shoot the foes? Most of the sensing infrastructure, notably satellites, is not expended but, rather, an investment. Expendable sensors ought to be cheap. Munitions, if bought in quantity, are also cheap; at the height of the 1980s buildup, PGM procurement accounted for only one of every 40 defense dollars. As for platforms and their units—the traditional metric of power—their job would be to haul sensors and weapons closer to the battlefield. Costs, as it turns out, would depend little on how many enemies exist, or how many tanks, planes, and soldiers enemies have. Instead, they would more likely depend on an adversary's sophistication and experience at deception; these two factors would indicate how much redundancy is needed in sensors and weapons.

Conventional War, Hyperwar, and Mud Warfare

Traditional warfare à la Desert Storm, hyperwar, and mud warfare place fundamentally different demands on military information systems because each has its own set of problems that information must be used to solve. To generalize, in traditional combat, what and (broadly speaking) where the enemy is are known. In this case, information systems need to generate data to provide situational awareness to operators. Complexity must be mastered to conduct large operations, and speed helps because it permits action before the adversary can react.

Conventional warfare pits superior force against force. The new logic of warfare, at least for the United States, is to scan the battlefield, sift through fields of data looking for targets, sort them by priority, and then strike those worth hitting. This approach has broad potential for stopping an approaching military and then wearing it down so that more-traditional combat approaches can be used to eventually push it back by successively occupying territory. In some cases, land occupation may not be necessary. If it is, successive attrition makes the job far easier—and, as noted, land forces need not wear U.S. uniforms.[5] Even in nonlinear combat, the ability to illuminate the battlefield can inhibit the enemy from massing forces and thereby from carrying out operations that require doing so. Since everyone makes mistakes, the ability to spot and exploit those mistakes quickly enough permits one to take advantage of them. It may be possible to know which locations the enemy is paying no attention to and thus where friendly forces can operate. Illumination allows supplies to be interdicted more successfully and permits the battlefield to be divided into smaller and smaller cells, each of which can then be attacked.

This capability is a goal and, as NATO operations in Kosovo imply, not yet a reality. Post-war analyses suggest that military assets were hidden in forests and villages and were not easily detected if they

---

[5]This presumes that the interests of the United States and its allies are sufficiently similar. Such is not always the case, as Operation Anaconda revealed. The United States was far more interested in destroying Al Qaeda than its Afghan allies were.

were not moving at the time. That noted, rendering them immobile might have been half the battle won. Unfortunately for the United States, it is precisely the prospects of such warfare that may compel adversaries to think of war in other terms—an example of how innovation for one can spark counterinnovation for others.

One response, termed mud warfare, is likely to be characterized by operating in dense environments, hiding in clutter, wielding civilian-looking material, taking hostages, and using terrorism. Mud warfighters have to learn what to look for and then where to look. Information plus training should prepare warfighters to distinguish patterns of hostile activity from everyday backgrounds. They must know where to intervene quickly when order is tipping over to chaos or enemy control. Because mud warfare features small unit operations, the locus of command is best moved down.

In hyperwar, warfighters know what to look for but not exactly where to find it. Complexity is associated with picking high-value targets from the clutter. Speed is necessary to track and engage such targets while they are visible. The availability of global information and the ability to bring any and all weapons within an ever-larger radius into simultaneous play tend to move command up.

Table 4.1 suggests how the three types of warfare affect C4ISR requirements. Paradoxically, while mud warfare appears crude, the information architecture required to conduct it is, in fact, much more complex than that for hyperwar. Such is the ability of future technology to shape competition that, in turn, reshapes the requirement for technology.

Table 4.1

C4ISR Requirements for Conventional War, Hyperwar, and Mud Warfare

	Conventional War	Hyperwar	Mud Warfare
Environment	Knowns	Known unknowns	Unknown unknowns
Purpose	Data/imagery	Information	Understanding
Complexity	Running ops	Finding things	Sensing patterns
Speed	Gain cycle-time edge	Find fleeting targets	Preempt tipping points
Command	Today's mix	Moves up	Moves down

FALSE REVOLUTIONS ON THE VIRTUAL BATTLEFIELD

If the real revolution in information technology lies not in its continual improvement but in the *form* that its improvement takes—distributing processing power into smaller packages and amalgamating it, in turn, into more powerful networks—does war follow commerce into cyberspace, pitting foes against one another for control of this clearly critical high ground? Does this comparison have a basis in tomorrow's reality, much less today's?

The Defense Science Board seems to believe it does:

> The objective of warfare waged against agriculturally-based societies was to gain control over their principal source of wealth: land. . . . The objective of war waged against industrially-based societies was to gain control over their principal source of all wealth: the means of production. . . . The objective of warfare to be waged against information-based societies is to gain control over the principal means for the sustenance of all wealth: the capacity for coordination of socio-economic inter-dependencies. Military campaigns will be organized to cripple the capacity of an information-based society to carry out its information-dependent enterprises.[6]

What Is Information Warfare?

The purpose of information is, was, and always will be to inform decisions; if not, it is just entertainment. Prior to World War II all these decisions were made by people. With the advent of digitized information systems, an increasing share of decisions—choices made among alternative actions—are made by machines. But they are decisions nonetheless.

Because conflict has always involved decisions, information has always been a part of conflict. Information warfare can thus be defined as the actions taken to influence the enemy's decisionmaking processes so that its decisions are bad or too late or good for your side (e.g., deciding to stand down rather than fight). Seen in this light,

---

[6]*Report of the Defense Science Board Task Force on Information Warfare—Defense,* Office of the Under Secretary of Defense for Acquisition and Technology, Washington, DC, November 1996, p. 2-1.

quintessential human activities such as deception, propaganda, and targeting the other side's commanders are hardly new to warfare.

What is new is that many of today's decisions are made using more information, and that information is richer, generally more timely, and often more widely disseminated. At first glance, this tendency supports a belief that information has become a center of gravity for military operations—and, as such, more deserving of attack. Attack methods, it would seem, now merit greater resources. At second glance, however, the logic is a bit odd. It is as if to say that the task of toppling someone sitting on a stool is more likely to call for breaking the stool's legs if they number eight rather than if they number three. Sound strategy requires looking not only at an opponent's *demand* for information but also at its *supply*. The latter is determined by long-term trends in technology, and these trends are not especially conducive to information warfare. The exception is computer warfare: because computers are everywhere and connected, defending them is harder.

To demonstrate as much requires that information warfare be decomposed. Consider the following information flow: Information is gathered by sensors, relayed to decisionmakers through the electromagnetic spectrum, and reaches the command center, where it is entered in a computer for further processing, the results of which inform decisionmakers. Figure 4.2 shows the primary related forms of information warfare: attacks on sensors, electronic warfare, C2 warfare, computer (hacker) warfare, and psychological operations. Note, again, that with the recent exception of the computer, and the half-century exception of the spectrum, none of this is particularly new, especially if sensors are considered today's version of yesterday's cavalry pickets.

What does the future tell us about the efficacy of information warfare? Modern militaries *are* increasing their dependence on the ability to generate and use information gathered and exploited from well over the horizon. But the industrial-era paradigm of concentrating value in a few machines of increasing complexity may have peaked and, with it, the vulnerability of modern systems, even military systems, to information warfare. Again, computer warfare is an exception, one dealt with below.

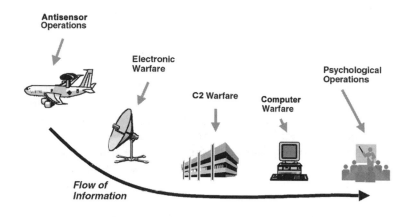

Figure 4.2—Five Types of Information Warfare

Antisensor Operations

The flow of good information into sensors can be disrupted in one of three general ways: (1) destroy the sensor itself, (2) use arcane electronic methods to spoof the sensors into seeing the wrong set of bits, and (3) use cover, concealment, and deception to fool those who interpret these bits into not seeing what is really there (or into seeing what is not there). The second method is specific to electronic equipment (notably radar-based equipment); the other two clearly predate electronics. The primary reason antisensor warfare may be more attractive today is that sensors are more important to warfare than ever before; the growing ability to hit targets at a distance depends on being able to detect and identify them from comparable distances.

The vulnerability of U.S. sensors to attack depends on the attacker. The primary long-range sensors in the U.S. inventory include reconnaissance satellites, aircraft (e.g., JSTARS, AWACS, Rivet Joint, Cobra Ball, the U-2), large radars (e.g., the Aegis system), and UAVs. So far, only Russia can attack a reconnaissance satellite (although China may soon have some capability). Similarly, aircraft and ships can be protected by operating 100 km or more behind the lines and using a protective screen (e.g., F-15s) to keep adversaries at bay.

Twenty years out, the survival of such assets may be dicier, especially against an opponent that determinedly invests to neutralize a decisive source of U.S. advantage. Reconnaissance satellites are expensive, few in number, and fly rather close to the earth. A spruced-up Scud-like rocket on direct ascent may be able to intercept their orbits, and lasers may be employed to blind at least the optical sensors on such spacecraft. Radar-based aircraft are large, hard to maneuver, easy to detect by watching for their energy output, and not at all stealthy. If they have to track deep targets or if front lines are poorly held or defined, they may be vulnerable, particularly to very long range missiles. The Aegis cruiser is at risk because naval fleets are developing increasingly sophisticated defenses against cruise missiles but usually lack the stores to completely counter a sufficiently large attack volley. UAVs are somewhat better protected due to their stealthy characteristics, their ability to fly high, and their small size.

UAVs, however, foretell the coming futility of direct antisensor operations. As electronics get cheaper, the prospect of putting formidable electro-optical capabilities on increasingly smaller and less-expensive UAVs suggests a long-term strategy of employing such sensors in large profusion. Sufficiently cheap but increasingly sensitive UAVs cost less to replace than to destroy. Finally, if unmanned ground sensors approach the low cost of consumer electronics (e.g., microphones, digital cameras) they can provide an even more robust method of collecting signatures.

The fate of electronic spoofing follows a similar logic. It is far easier to spoof a single sensor of known phenomenology (e.g., radar, electro-optical, hyperspectral, passive millimeter wave, acoustic) than to fool a suite of sensors that are in multiple locations and use various phenomenologies to produce readings that can be correlated to develop an increasingly detailed view of the battlespace.

This leaves the opponent with the age-old techniques of making the hostile appear benign (e.g., putting troops in vehicles normally assigned to hospitals) and feeding the preconceptions of adversaries so as to persuade them to classify the normal as threatening. These techniques, in large part, predate technology.

Electronic Warfare

Can messages be stopped, corrupted, or intercepted between nodes in the system? Jamming, message spoofing, and interception—all were the stuff of the so-called Wizard Wars conducted between Britain and Germany in the early 1940s.

One form of jamming works by putting an emitter between two enemy communications devices so that the noise blocks the message. As a general rule, the cost of transmitting a signal of given strength (being largely a matter of power chemistry) may not change very fast. However, the adroit use of digital waveforms, the ability to exploit spread-spectrum and frequency-hopping to hide a narrowband message within a wideband slice of spectrum, and the reduced cost of making receivers are lengthening the edge enjoyed by message senders over message blockers. Tomorrow's phased-array receivers should be able to focus on a transmitter of known location and tune out interference from the side. Thus, if the jammer does not sit between the receiver and the transmitter, it can get in the way only by generating far more power. Phased-array receivers may well become commercial items (the increasingly scarce spectrum can be reused if receivers can be pointed at a specific transmitting antenna to the exclusion of others around it).

Radar jamming still has some life, however, because the source of the signal, as it were, is the reflection off the target, near which the jammer can fly, sail, or stand. The nearness of jammers and targets obviates the line-of-sight defense available to communicators. Radars may instead emit constantly changing pulsed digital waveforms and collect an electronically unique signature, coupling that with mathematical techniques that help separate signal from noise.

As for spoofing and interception, any message as long as a short paragraph may be protected with cryptographic techniques. Transmitters equipped with a private digital key can add a signature tag to a message; this tag can authenticate the transmitter's identity and the fact that the message has not been corrupted.[7] (If the message is

---

[7]The message is mapped into a unique 128-bit hash, which is then processed together with the private key to form a signed block. The receiver takes the public key, reverse-processes the block, extracts the hash, and checks for consistency between the

time-stamped, it cannot be echoed deceptively, either.) Encryption is a well-known defense against eavesdropping. Until recently, encoding everything would have been computationally burdensome, but cheap electronics have ended that problem. Barring some unforeseen mathematical breakthrough or the development of quantum computers of sufficient size (unlikely before 2020), the time required (even by a palmtop) to encode a message will fall—regardless of whether corresponding advances in supercomputers require longer keys to preserve message confidentiality.

In sum, while electronic warfare has always been a seesaw affair, developments over the next 20 years will continue to favor the bits getting through and will generally favor their being read.

Command-and-Control Warfare

The third target for information warfare is the command center itself, using shot and shell to disable the commanders, destroy the command apparatus, or sever the wires and fibers to fielded forces. Electronics are also vulnerable to such soft-kill techniques as microwave bursts and electromagnetic pulses. Apart from the obvious ability to put a precision weapon anywhere within an identified command center, the really new feature of C2 warfare is the modern military's dependence on keeping information systems going—a dependence whose importance, in some respects, even surpasses the importance of keeping individual commanders safe from harm.

But here, again, the vulnerability of command centers may have already peaked because of the proliferation of digital electronics. The traditional architecture of information systems involved large, complex central office switches and mainframe computers. But trends everywhere are toward dispersion. Central office switches are being replaced by routers of decreasing size and increasing number; fixed routing is being replaced by packet switching, thereby permitting every packet to take a different route. Network trees can be replaced by network meshes. Large computers are being replaced by conglom-

---

message and the hash. A time-stamp confounds problems that could arise if the interceptor were to echo the original message back. Similar techniques using private and public keys can be used to exchange other keys that permit more-efficient and harder-to-break symmetric code breaking to be performed.

erations of smaller ones. Fixed storage is being replaced by redundant arrays of independent disks capable of being distributed throughout a network. From late 1993 to late 2002, for instance, the cost of hard storage fell by a factor of 1,000, a drop roughly twice as fast as that for the cost of processors (as measured in dollars per instruction/second). Although the cost of laying a line of fiber is unlikely to decline much, the capacity of that fiber is nearly limitless. If redundant lines are paid for and used, only one line need survive to support all the traffic needed, as long as increasingly sophisticated routing algorithms can redirect traffic at a moment's notice. Finally, dispersed power sources, such as photovoltaics and fuel cells, promise to make information systems at least somewhat independent from the still-vulnerable power grid.

Technology even offers a way to protect the command hierarchy. Given the growing realism of videoconferencing, the utility of whiteboarding[8] tools, and shared access to databases, people no longer need to be physically drawn together for command conferences. This limits the vulnerability of a force's leadership to a lucky strike. And so, the bottom line remains the same. Whereas modern militaries are increasingly dependent on information systems, the technological advances that have made dependence so attractive are also available to protect such systems with increasing confidence.

Psychological Operations

What of decisionmakers themselves? Here the ancient truism comes to the fore: to make adversaries yield, it helps if they are convinced that the benefits of cooperation are high, the cost of resistance is destruction, and that they are operating against the will of the heavens. As a general rule, the dependence of militaries on human factors is no greater than it always has been.

Psychological operations may be defined as the use of information to affect human decisionmaking. They range from attempts to influence the national will, to deception against the opposite commander, to propaganda against opposing forces. The broader term, *psy-*

---

[8]Whiteboarding is the ability to put a word, picture, etc., on a screen that then appears on all screens of all people in a session.

chological warfare, refers to all aspects of combat that affect the willingness of people to fight above and beyond any physical harm that befalls them. Ultimately, however, there is no form of warfare that is not, to one degree or another, psychological warfare.

Some aspects of psychological operations perforce vary with specific circumstances and power relationships (e.g., among politicians, military officers, and military forces). Other aspects reflect pervasive trends. One such trend is the growing openness of societies to external influences—from cable television, direct broadcast television and radio, and the Internet. Over the next 10 years, the proliferation of space-based multimedia broadcast satellites, the continual spread of fiber optic lines, decent language-translation software, the proliferation of remote sensing satellites in commercial hands, and the possibility of agents and bots that can transfer satisfying answers to vaguely worded questions should increase permeability much further.

Will technology make psychological operations more rewarding? Citizens, soldiers, and commanders will be subject to a vast array of data sources, making it harder to cut them off from what the world is saying and to cut the world off from what they are saying and seeing. Further, media dominance by a handful of major networks is moving toward an era of 500-channel television and into Me-TV, where the low cost of creating a video feed and the ability to mix and match sources mean that everyone's news sources are different. The more the potential sources of information cover any one incident (think the random videotaping of the Rodney King incident), the harder it becomes to put out a single effective story line.

The saving grace for the information warrior who wishes to deceive others is still the human tendency to jump to conclusions, willingly consuming supporting evidence, and filtering out everything else. A case in point is Hitler's certainty that the Allies would invade Europe at Calais. Here, too, the evolution of information technology is of little help to tomorrow's information warrior.

The Ghost in the Machine

The most insidious form of information warfare, and the one that has garnered the most media and high-level policy attention, is the ability to get inside information systems in order to render them dysfunctional. If subverted in this way, an information system may work poorly or collapse, permit the entry of corrupted information, or reveal its secrets. Possible hacker entryways include surreptitiously inserting code into the machine at birth (e.g., through deliberately queered circuitry) or later (e.g., viruses), and assuming the identity of an authorized user—or, better yet, a systems administrator—and issuing malicious commands.

For this type of information warfare, cost factors that lead to a proliferation of nodes work against defenders. The more nodes on a system, the more doors for a hacker to try and the more difficult to find the hacker's entryway and where the corruption actually lies. The ability of one node to transfer instructions or control to another permits, among other things, very agile routing around damage, but it also permits vicious viruses, bad bots, and aggravating agents to spread quickly throughout a system.

Clearly, therefore, hacker warfare is more likely to be effective in at least the near future than it has been in the past. But does that mean that hacker warfare can be anything more than a minor annoyance or the source of random damage? That was the case for the April 1999 Chernobyl virus and the October 2001 Code Red Worm, even though both caused hundreds of millions of dollars of damage.

In gauging the effectiveness of future hacker attacks, it is useful to classify information systems into castles and agoras. *Castles* are a nation's critical infrastructures—military C4ISR systems, funds transfer, safety regulation, power plants and similar industrial facilities, telecommunications switching systems, and energy and transportation control points. They are, or should be, generally self-contained units, access to which can and ought to be restricted. *Agoras* are the great consumer marketplaces of cyberspace, in which increased vulnerability to malice, accident, and dysfunction is the price paid for the dense interactions and potential learning experiences that contact with strangers permits. It is as hazardous to use the rules of the agora to govern the castle as it is constricting to enforce the castle's norms on the agora.

In the short run, predicting the course of information security almost requires predicting where the next set of mistakes will be made.

Complexity seems to swing the advantage toward the hackers, who know that finding just a single breach may open the floodgates. But complex systems need not necessarily be insecure. You, the readers, who are several orders of magnitude more complex than any man-made system, are proof of that. No combination of bits that you can read right now could wreak havoc in your operating system. I, the author, have no authority to make you do stupid things. You process these words not as instructions to be obeyed but as data to be ana-lyzed.

Processing inputs as information is the key here. As systems grow more sophisticated, they are likely to become more humanlike. They will be able to absorb data from beyond themselves, filter those data, and analyze them with a sophistication that grows with everything learned—just as humans do. True, critical castle systems can and probably should still be isolated. But the agora systems are fair game. And so, information systems will be heir to the more subtle faults. A computer that analyzes intelligence on a country, for instance, may absorb the content of Web-based newspapers, police reports, crop statistics, tax records, and local bulletin boards to draw conclusions.

These conclusions may, for their part, be influenced by what oth-ers—often self-serving and sometimes hostile—post, and computers will have to filter out the chaff to get the nuggets. Learning systems, such as neural nets and knowledge engineering devices, may be cor-rupted by bad information introduced at an unknown time, the dis-covery of which may leave administrators wondering how much good learning has to be erased to remove the bad learning. But this is essentially no different from what humans do, and, so far, humans have coped. The result is that hacker activity will express itself not as looming catastrophe, but as the certainty of at least some level of pollution. And pollution is not warfare.

THE LESSON OF SEPTEMBER 11

Peter Schwartz, founder of the Global Business Network, once ob-served that even futurists consistently underestimate the effects of technological inputs on the course of history.[9] What is underestimated is not the extent of change—i.e., seeing 1, forecasting 2, and realizing 4—but its breadth and depth. Were the problem a question of extent, the fix would be easy: double everything. But insufficient breadth and depth come from not seeing the variety of change, a failure that an overactive imagination cannot easily fix.

The events of September 11 remind us of the inevitability of surprise. Suicide bombers, the use of airplanes to attack symbolic monuments,[10] the anti-American sentiments held by fundamentalist Muslims, and the special animus against the World Trade Center—none of these was particularly new on that day. One prominent futurist who asked what the worst thing a terrorist group could do to America was, was repeatedly told by security experts that it could crash a 747 into the World Trade Center. He thus learned to dismiss this scenario as a cliché. But the event, itself, was still a surprise. Had something of this magnitude taken place in cyberspace rather than in real space, several information warfare gurus would have proclaimed, "I told you so." But they are not right, yet.

The success of the U.S. campaign against the Taliban regime validated, at least in that context, the hopes of believers in the current RMA. While neither the Taliban nor Al Qaeda is assuredly destroyed for all time, precision warfare, carried out in concert with willing if initially less-powerful allies, did in three months what the Soviet Union could not complete in 10 years.

---

[9]Peter Schwartz, *The Art of the Long View: Planning for the Future in an Uncertain World,* Doubleday, New York, 1996.

[10]Several years earlier, French security forces forestalled a plot to launch a jet into the Eiffel Tower.

Chapter Five

UNCERTAINTY-SENSITIVE PLANNING

Paul K. Davis

Consider some of the major strategic surprises that affected national security in past decades.[1] Some of these were negative; some were positive:

- Cuban missile crisis

- Sadat's peace mission to Israel

- Fall of the Shah of Iran and resulting hostage crisis

- Disintegration of the Soviet Union

- Peaceful reunification of Germany

- Iraq's invasion of Kuwait

- East Asian economic collapse of the late 1990s

- India's nuclear testing and Pakistan's response

- Terrorist attacks on the World Trade Center and Pentagon

Now consider some purely military surprises of the past 50 years:

- Torpedoes in U.S. submarines fail to detonate (World War II)

- Early air-to-air radar-guided missiles fail (Vietnam War)

- Egypt launches a surprise attack across the Suez canal (1973)

---

[1]The intelligence community provided what is called "strategic warning" in a few instances (i.e., warning days or weeks before the events).

- Israel's air force is stymied by surprisingly effective SA-6 batteries

- Israel achieves an astounding exchange ratio in air war over Lebanon's Bekka Valley

- Deployment to the Persian Gulf begins a week *after* the attack (Desert Storm, 1990)

These surprises form a long list because they have not been occasional annoyances in a generally predictable world but, rather, quite common. This chapter begins by discussing why addressing uncertainty is so important and difficult. It then discusses analytic methods for dealing with uncertainty in broad, conceptual strategic planning; in doing capability-based analysis of future forces; and in making choices within a budget. Some of these methods are now in use in the Pentagon and elsewhere; others are candidates for future implementation. Resisting the tendency to give short shrift to uncertainty requires both discipline and knowledge of the past. Conscious methodology can help provide some of that discipline.

WHY SO MANY SURPRISES?

Early in World War II, U.S. submarine commanders were horrified as their torpedoes—launched after dangerous approaches—passed harmlessly by their targets. The torpedoes had worked in laboratory testing; their failure in the field was so unexpected that the commanders' reports were initially not believed. In the Vietnam War, early radar-guided air-to-air missiles failed because engagements occurred in circumstances other than those planned for. Faith in the missiles had been so great that most aircraft were designed without backup gun systems. Both of these failures were, in a sense, "merely" technical and analytic, but they had major consequences for the prosecution of war.

Nor have historical surprises all been American. When Egyptian units seized the east bank of the Suez Canal early in the 1973 Yom-Kippur war, the Israeli Air Force was stymied by surprisingly effective SA-6 anti-aircraft batteries. In later years, the Israeli Air Force destroyed scores of Syrian aircraft over the Bekka Valley without losing a single aircraft, thanks to asymmetrically capable command and control (C2). This may not have been a surprise to the Israelis, but it certainly got the attention of militaries worldwide, including that of the Soviet Union, which supplied the Syrians.

In more modern times, there have been different surprises. Even though official U.S. planning scenarios had almost always assumed that U.S. forces would be able to deploy well before the day war would begin, or D-Day, the Desert Storm deployment began six days after D-Day. More recently, in the war over Kosovo, NATO heads of state were reportedly surprised when Slobodan Milosevic was not brought to his knees immediately by NATO's bombing. Their confidence was so great that they prohibited the development of contingency plans involving ground-force operations until late in the campaign. To make things worse, the Yugoslavs did not "play fair," usually keeping their air-defense radars turned off. Destruction of their air defenses thus proved impossible, constraining operations, reducing their effectiveness, and increasing the mass of air forces needed. Serbian military forces dispersed in the woods and emerged from the war with little damage. The United States was badly stretched while prosecuting a one-sided war against a third-rate regional rogue. Ramifications of this unscheduled "small-scale contingency" echoed throughout the entire force structure (particularly for the Air Force, which bore a particularly heavy burden in this conflict).[2]

Why do so many predictions fail and surprises occur? The reasons include the constant competition of measures and countermeasures, the tendency to keep weaknesses out of mind only to have them attacked by the adversary, rather prosaic failures of design or execution, and a failure to appreciate the frictions of war celebrated in the 1832 writings of Clausewitz.[3]

The scientific way to look at uncertainty is to acknowledge that wars and military competitions are "complex adaptive systems," and that, as a result, even small events can and often do have large effects. Further, the "system" is not a constant for which one can prepare

---

[2]As usual, not everyone was equally surprised. Some military leaders were pessimistic from the outset about the effectiveness of bombing conducted with the severe political constraints that applied in the early weeks of the campaign.

[3]Carl von Clausewitz, On War, translated by Peter Paret, Everyman's Library, Alfred Knopf, New York, 1993.

straightforwardly. Rather, it includes human beings and organizations that think, behave, and adapt to events in myriad ways.[4] Because of such complications, an accurate prediction of the course of events is sometimes not even feasible.[5] That is, uncertainty is not only ubiquitous and large, but also impossible to get rid of it by merely working hard to do so.

So what do we do about this burden of uncertainty? Do we just wring our hands? In a phrase, we should get on with business—learning to plan in a way that includes the expectation of surprises and the need for adaptations. Until recently, this admonition seemed radical to defense planners, but it is old hat in many other endeavors, ranging from professional sports to U.S. business.[6] It is also quite familiar to warfighters.

CONCEPTUAL STRATEGIC PLANNING

Uncertainty-Sensitive Strategic Planning

Strategic planning can be expensive, tedious, and counterproductive or lean, stimulating, and insightful.[7] The uncertainty-sensitive

---

[4]For an excellent semipopular discussion of complex adaptive systems, see John Holland, *Hidden Order: How Adaptation Builds Complexity*, Addison Wesley, Reading, MA, 1995.

[5]The degree of unpredictability depends on circumstances and what is being predicted. If a massive opponent attacks a much smaller opponent in an "open field," the outcome is determined. When dealing with complex adaptive systems, a key is knowing the "envelope of circumstances" within which control is possible. For a sample discussion on the control of nonlinear complex adaptive systems, see Appendix B of National Research Council, *Modeling and Simulation*, Vol. 9 of *Tactics and Technology for the United States Navy and Marine Corps, 2010–2035*, National Academy Press, Washington, DC, 1997.

[6]For earlier discussions that seemed more radical at the time, see Paul K. Davis (ed.), *New Challenges for Defense Planning: Rethinking How Much Is Enough*, MR-400-RC, RAND, 1994, Chapter 3, and Paul K. Davis, David C. Gompert, and Richard L. Kugler, *Adaptiveness in National Defense: The Basis of a New Framework*, IP-155, RAND, 1996. Planning for adaptiveness is now well accepted by Department of Defense (DoD) leadership (Donald Rumsfeld, *Report of the Quadrennial Review*, Department of Defense, Washington, DC, 2001).

[7]For an excellent but caustic review of failed strategic planning efforts, mostly in the business world, see Henry Mintzberg, *The Rise and Fall of Strategic Planning*, The Free Press, New York, 1994. Some of Mintzberg's criticisms of elderly strategic planning processes apply well to DoD's planning, programming, and budgeting process (PPBS).

planning method is designed for taking an occasional fresh look at the future's challenges and possible strategies—for rethinking matters such as grand strategy and higher-level defense planning. Variants of this method have been used at RAND since the late 1980s, as the Cold War was ending.[8]

Figure 5.1 indicates the basic ideas of uncertainty-sensitive planning. The first step is to characterize the "core" environment, sometimes called the *no-surprises future*. The next step is to identify uncertainties of two types related to *branches* and *shocks*.

Branches represent uncertainties that are taken seriously and monitored and that will be resolved at some point once events take us down one path rather than another. These uncertainties can be dealt

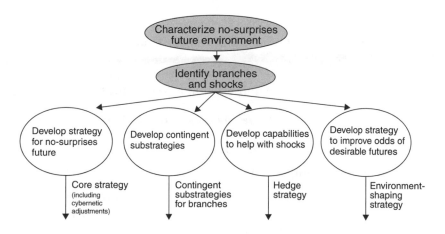

Figure 5.1—Uncertainty-Sensitive Planning

For an informal survey that discusses more strategic methods than can be mentioned here, see Paul K. Davis and Zalmay Khalilzad, *A Composite Approach to Air Force Planning*, MR-787-AF, RAND, 1996.

[8]For more details, see Davis, *New Challenges*, Chapter 6, which summarizes work done some years earlier in collaboration with Paul Bracken. James A. Dewar et al. (*Assumption-Based Planning: A Planning Tool for Very Uncertain Times*, MR-114-A, RAND, 1993) articulate well a variant method, assumption-based planning, that is especially useful for critical reviews of existing plans. It has been refined and applied extensively and James Dewar has recently published a book on the subject (James A. Dewar, *Assumption-Based Planning*, Cambridge University Press, Cambridge, UK, 2002).

with by in-depth contingency plans. Shocks, in contrast, involve plausible (i.e., not impossible) events that are heavily discounted by best-estimate wisdom and are given lip service, at best. Nonetheless, at least some of them will occur—even if they are individually unlikely. When they do occur, they will be disruptive, and there will be no detailed contingency plans for dealing with them. As suggested by the examples listed at the beginning of the chapter, such events not only occur, but occur frequently.

As of early 2002, some illustrative future branches for current U.S. strategic planning included whether

- Al Qaeda is eradicated.
- The U.S. takes military action to force a regime change in Iraq.
- Korea is unified.
- China engages in military actions against Taiwan.
- NATO expands to the Baltics.
- The long-term Chinese military buildup continues.

Some of the many shocks currently regarded as unlikely might be

- U.S.-China conflict arising from a Chinese attack on Taiwan.
- Revolution in Saudi Arabia.
- Collapse of Iran's Islamist government and movement toward normalization (or the opposite, a resurgence of virulent anti-American Islamist-driven actions).
- Disintegration of extremist Al Qaeda-like movements.
- Resolution of the Arab-Israeli conflict and emergence of a Palestinian state.
- Japan going independent.
- Russia moving against Baltic states, Ukraine, or Poland.

As Figure 5.1 shows, planners are to develop a broad strategy consisting of the no-surprises strategy, a series of contingent substrategies to deal with branches, a set of hedging actions laying the groundwork for more ad hoc adaptation to shocks when they occur, and an environment-shaping strategy to affect favorably the odds of various futures. Three particular themes are crucial here:

1. *Operational adaptiveness* is the ability of U.S. forces to deal, at a given time, with a diversity of political-military scenarios and detailed circumstances—some of which can be planned against in detail and some of which will arise as shocks.

2. *Strategic adaptiveness* is the ability to change military posture quickly and easily in response to shifts of the geostrategic environment or national strategy. "Quickly" relates to the time scale of changes in the environment (years); "easily" relates to budgets and effectiveness. Again, some possible shifts can be anticipated and planned against; others will be surprises.

3. *Environment shaping* is influencing the future—e.g., by promoting international stability, economic integration, and universal democratic principles; controlling or mitigating international instabilities; and underwriting general deterrence through commitments, relationships, and credible military forces.

Planning for adaptiveness is more easily said than done, but the United States appears to have often been rather good at it when viewed through the lens of what has been called effective "muddling through."[9] In the real world, the best that can be done is to move in the "right direction" and to adapt routinely without falling prey to the illusion that more precise planning is possible.

Environment shaping is perhaps best understood by considering its opposite: treating the future as an exogenous variable over which one has no influence. Doing so is common in strategic planning activities that spin alternative futures.[10] But the future obviously is not exogenous, or given—especially for a superpower. Humility about shaping efforts is one thing (such efforts can surely fail or be counterproductive), but just waiting to see what happens is quite another.

---

[9]Charles Lindbloom, "On the Science of Muddling Through," *Public Administration Review*, Spring, 1959.

[10]A caricature here is a scenario-based approach that includes a rosy scenario, a bad scenario, and a no-surprises scenario. Sometimes participants emerge with few insights other than that they prefer the rosy scenario.

Surely, the United States should seek ways to improve the odds of favorable developments and circumstances.

Shaping has a positive side, such as seeking to expand the zone of peace; it also has a side that forestalls the negative, as in establishing general deterrence in a given region. It is here that environment shaping connects with the classic strategic concepts of *realpolitik*. Even an optimist about the arrow of human progress has to recognize that military vacuums do arise, malevolent leaders exploit them, and wars still occur. It is far better to deter such events by maintaining a manifest capability to deal with them should they occur, than to have to actually fight future wars. The shaping concept has become increasingly important over the last decade. It seems now to be well established in U.S. national strategy, although terminology changes with each administration, as does the relative emphasis on carrots and sticks.[11] As of late 2002, the United States was considering preventive war against Iraq. Such a war would likely have major longer-term shaping effects, for good or ill.

Operationalizing Strategic Planning in Portfolio-Management Terms

Conceptual strategic planning, then, can address uncertainty in the way suggested in Figure 5.1. The next issue, however, is how to move from that to something more formal, structured, and actionable— i.e., to seeing defense planning as an exercise in portfolio manage-

---

[11]The first official embrace of the "environment shaping" idea was that of Secretary of Defense Dick Cheney, in 1993 (Dick Cheney, *The Regional Defense Strategy*, Department of Defense, Washington, DC, 1993). The early Clinton administration dropped the terminology, but embraced the related concept of engagement. The concept of preventive defense (Ashton B. Carter and William J. Perry, *Preventive Defense: A New Security Strategy for America*, Brookings Institution, 2000) is about certain types of environment shaping. The broad concept of environment shaping per se was reintroduced to official documents in 1997 (William S. Cohen, *Report of the Quadrennial Review*, Department of Defense, Washington, DC, 1997). For a region-by-region discussion of what environment shaping may involve, see Institute for National Strategic Studies (INSS), *Strategic Assessment 1998*, National Defense University, Washington, DC, 1998. The Bush administration referred early to a strategy of dissuading, deterring, and defeating enemies and of reassuring allies (Rumsfeld, *Report of the Quadrennial Review*). The implications of such a strategy clearly include what is called environment shaping here.

ment. The intention of this construct, when first proposed in 1996 as background for the 1997 Quadrennial Defense Review (QDR), was to

- Promote capabilities-based planning for diverse contingencies, both large and small.

- Give environment shaping and strategic adaptiveness the same visibility and status as warfighting.

- Emphasize the need for hedge capabilities permitting future adaptiveness.

- Deal with the potential synergy of and conflicts among portfolio components.

In this construct, planning is about judging how best to allocate investments across the three components of the portfolio. Figure 5.2 suggests schematically what this can mean.[12] The left component

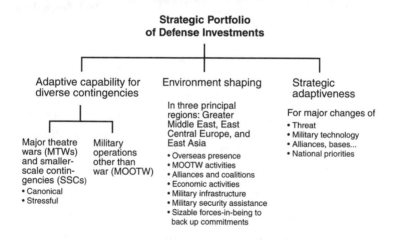

Figure 5.2—Defense Planning in a Portfolio-Management Framework

---

[12]Adapted from Davis, Gompert, and Kugler, *Adaptiveness in National Defense*, which also discusses similarities and differences between this type of portfolio management and that of the business world.

highlights capabilities planning, since that is DoD's core mission; the subsequent components deal with environment shaping and hedging activities to achieve strategic adaptiveness designed to prepare for an uncertain future.

Portfolio management has its limits. As in the investment world, actually doing portfolio management is by no means straightforward, but it is a coherent way to deal with inherent uncertainty and multiple objectives. Choices must be made because there are conflicts. Maintaining near-term readiness can conflict with building future-year capabilities. Worldwide shaping activities can shortchange modernization and transformation of U.S. military forces. Overzealous transformation efforts can mean low readiness until forces, doctrine, and the personnel system adjust.

Subtleties also abound. For example, many military capabilities (first component in Figure 5.2) add or subtract from environment shaping. Thus, environment-shaping investments are investments to *further* increase U.S. effectiveness beyond that stemming from capabilities. Similarly, many activities to enhance both conflict capabilities and environment shaping can add or detract from strategic adaptiveness (e.g., by creating options for using nations' ports and airfields in the future as others become unavailable or undesirable).

Another subtlety is that merely labeling some activity with a positive, such as "increased presence improves environment shaping" may confuse intent with reality. A forward presence can have negative effects when it is too intrusive and runs afoul of independence, sovereignty, or pride. So it was that the United States lost its base in the Philippines. Similar problems have arisen in Korea, Okinawa, and the Persian Gulf. It would not be surprising if the U.S. presence in Saudi Arabia were significantly reduced in the years ahead.

As Chu and Berstein observe in Chapter One, the DoD has, in the course of the last three administrations, substantially altered its concept of higher-level strategy to the needs of the modern era. This has often been obscured in debates over hearty perennials, such as whether to cut forces, cancel high-visibility weapons systems, or drop the requirement that the United States must be able to fight and win two major theater wars (MTWs) simultaneously. From the perspective of this chapter, however, a great deal of progress has been made. DoD's (and the nation's) strategy is explicitly multifaceted, as discussed here in portfolio-management terms. Attention is paid separately to current operations, recapitalization, transformation, and environment shaping. Force needs are assessed not in terms of a simplistic warfighting formula, but in terms of diverse worldwide requirements. There should be no doubt, then, about DoD's attention to the special challenges of planning under manifest uncertainty.

At the heart of those changes is the shift to capabilities-based planning of future forces, referred to above as DoD's core mission (and shown at the left in Figure 5.2).

CAPABILITIES-BASED PLANNING

Capabilities-based planning is planning, under uncertainty, that aims to provide capabilities suitable for a wide range of future challenges and circumstances while working within an economic framework. Today's defense planning, then, is about building capabilities that will be available perhaps three to 20 years from now, when future presidents, defense secretaries, and combatant commanders face the challenges of their own eras. Only sometimes will those challenges have been anticipated, and even less often will they have been planned for in detail despite all the trying. This implies that the capabilities provided to those future leaders should be designed for flexibility and adaptiveness.

Capabilities-based planning stands in contrast to what had become DoD's approach to planning, an approach based on official planning scenarios for major theater wars that not only identified adversaries, but also laid out scenario details, such as warning time and roles of allies. Figure 5.3 shows schematically the kind of scenario used. DoD's routine analysis processes had become so focused on these official scenarios, along with official databases for running official models, that the result was the virtual opposite of capabilities-based planning. It was as though the illustrative scenarios had become specifications serving to define both necessary and sufficient characteristics of the force structure. In practice, as so often happens when strategic planning processes age, the constrained analyses consistently supported current programs—i.e., they "caused no trouble."

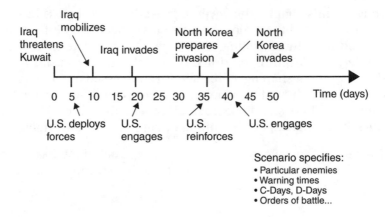

Figure 5.3—An Illustrative Threat-Based Scenario

They were not, however, useful for dealing with uncertainty or for assessing transformation concepts.[13]

It is important to emphasize that the problem with the approach was not that it identified particular threats, but that it considered only conventional-wisdom threats and, to make things worse, considered only point versions of detailed scenarios, as though the circumstances of future conflict could be predicted.

Key Features

The essence of capabilities-based planning is to deal with future uncertainty by generating "capabilities" usable for different purposes and circumstances. Its key features are

- An emphasis on modular (building-block) capabilities usable in many ways

- Assembly capability

---

[13]For an extensive discussion of capabilities-based planning and suggestions for its implementation, see Paul K. Davis, *Analytic Architecture for Capabilities-Based Planning, Mission-System Analysis, and Transformation*, MR-1513-OSD, RAND, 2002. Appendix A of that document lists numerous past examples of capabilities-based planning, mostly from the 1960s and 1980s.

- Goals of flexibility, adaptiveness, and robustness, rather than "optimization"
- Multiple measures of effectiveness (MOEs)
- Explicit role for judgments and qualitative assessments
- Economics of choice
- Recognition that "requirements" are the result of high-level choices that should be based on broad capabilities-based analysis.

Building blocks are central in capabilities analysis. When developing capabilities, one quickly discovers the importance of modularity: of having the capacity to take a bit of this and a bit of that, and to do something for which one had not previously planned explicitly. This approach is familiar in everyday life. Suppliers to builders, for example, do not stockpile materials fine-tuned to particular homes that may be built; rather, they stockpile bricks, mortar, and studs.

Building blocks in the military domain come in many forms and at many levels. In particular, there are multiple levels of building blocks in four dimensions:

- Units (e.g., battalions)
- Operations (or missions) and related suboperations
- Weapons systems and subsystems
- Support structures (e.g., logistics systems and, within them, individual systems such as prepositioning ships or tactical airlift).

Flexibility, adaptiveness, and robustness depend on skills in assembling building blocks for at-the-time purposes and circumstances. They are undercut by overspecialized acquisition, by not achieving the interoperability that allows the blocks to fit together easily, and by refining detailed operations plans rather than honing skills for rapid at-the-time assembly. Part of the assembly challenge is having the capacity for at-the-time tailoring—e.g., creating special hybrid units and unique types of support, rather than using only large, pre-existing support structures.

The U.S. military does capabilities-based planning now at lower levels—e.g., military systems typically have specifications assuring usability in a wide variety of conditions. Similarly, lower-level operations planning tends to be quite adaptive (that is part of what makes being a young officer attractive). In contrast, higher-level operations planning is often ponderous, especially in peacetime. Although Desert Shield succeeded, the original plan had many shortcomings, and rapid plan modifications were difficult to make. Fortunately, the United States had six months before its shooting war began.

Information Technology and Mission-System Analysis

A critical issue in capabilities planning is assuring that assembly of the right capabilities can be accomplished quickly and can draw on resources that may be physically distant and that come in variously labeled packages—e.g., Army, Air Force, Navy, and Marines. At least as challenging is the task of ensuring that theater commanders will be able to draw on civilian expertise back in the United States or elsewhere when needed and to coordinate activities with those of coalition partners and nongovernmental organizations (NGOs). All of this will require excellence in the use of modern information technology, as Martin Libicki stresses in Chapter Four.

A related issue is mission-system analysis, which seeks to assure that *all* components necessary to an operation will be successful—e.g., that there will be immediately effective joint command-control; missile defense; defense suppression; reconnaissance and targeting; fire delivery; and orchestration of ground force maneuvers. Such matters are extremely important and are becoming more so.[14]

---

[14]See National Research Council, *Network Centric Naval Operations: Transitional Strategy for Emerging Operational Capabilities*, Naval Studies Board study, National Academy Press, Washington, DC, 2000; David S. Alberts, John J. Garstka, and Frederick P. Stein, *Network Centric Warfare: Developing and Leveraging Information Technology*, CCRP Series of National Defense University, Washington, DC, 1999; Paul K. Davis, James H. Bigelow, and Jimmie McEver, *Analytical Methods for Studies and Experiments on "Transforming the Force,"* DB-278-OSD, RAND, 1999; David Gompert and Irving Lachow, *Transforming U.S. Forces: Lessons from the Wider Revolution*, IP-193, RAND, 2000; and Davis, *Analytic Architecture*.

Multiple Objectives and Measures

When making assessments in capabilities analysis, multiple objectives are customary. That may seem straightforward, but defense analysis too often focuses instead on what amounts to a single objective. It is worth illustrating why this matters.

Figure 5.4 shows two ways of comparing four options: A, B, C, and D. The left panel compares them on the basis of least cost for equal effectiveness for a single measure of effectiveness, MOE 1. Perhaps this is something like the ground lost in a simulated war for a particular detailed scenario. In this comparison, option C looks best. In contrast, the right panel makes equal-cost comparisons using a range of measures (MOEs 1, 2, 3, and 4). Perhaps MOE 2 represents results for a different scenario, MOE 3 represents results for an entirely different mission, and MOE 4 assesses the relative extent of U.S. losses that would be expected over a variety of missions and scenarios. Option C may still be best with respect to MOE 1, but not by very much. As a result, its score (as indicated by shading, with lighter being better) is essentially the same as that of the other options. More importantly, option C is distinctly inferior to option D under the other measures. Overall, then, it would probably be better to choose option D. This, in miniature, is what it means to focus on flexibility and robustness rather than on optimizing (e.g., minimizing cost) for a point problem.

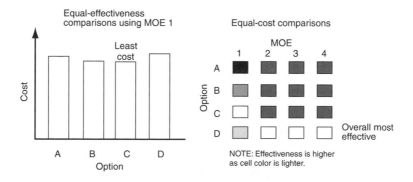

Figure 5.4—Contrasting Types of Analysis

The Concept of a Scenario Space (an Assumptions Space)

Having multiple MOEs can improve flexibility, adaptiveness, and robustness. Yet we must also recognize how sensitive force effectiveness is to highly uncertain variables often treated as if their values were known. Figure 5.5 shows how defense planning could be broadened, in two steps, from one or two threat-based point scenarios. The first step is to expand the list of name-level scenarios (i.e., scenarios specified only to the extent of giving their name, as in "China invades a unified Korea"). Recognizing this need for scope, DoD began in the mid-1990s to consider a broad range of generic threats, although most public attention still focused on threats from Iraq and North Korea.

What has not yet systematically occurred is the second step shown in Figure 5.5: for each name-level scenario, evaluate capabilities for a broad range of operational circumstances that would stress capabilities in very different ways. Ultimately, capabilities assessment requires an examination of outcomes for the entire *scenario space* (i.e., for all the combinations of factors that matter or for a truly representative distribution). This is *exploratory analysis under uncertainty.*

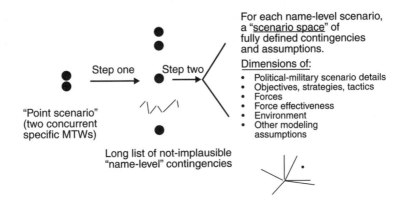

Figure 5.5—Exploratory Analysis in a Scenario Space

As the right side of Figure 5.5 indicates, it is useful to formalize this construct by recognizing that these factors can be grouped into six dimensions of scenario space:

- Political-military scenario (what is usually meant by "scenario")

- Objectives, strategies, and tactics

- Forces

- Force effectiveness (taking into account frictions and personnel quality)

- Environment

- Other modeling assumptions (e.g., feasible rates of advance and break points).

For every scenario space, an exploratory analysis covering the key uncertainties can be designed, and computers can then be used to "fly through the outcome space" to see under what assumptions war outcomes would be favorable, unfavorable, or uncertain. The purpose is to gain insight about how to avoid being in the bad regions of scenario space.[15]

To conduct an exploratory analysis, one can use the scenario-space dimensions explicitly, designing experiments to cover combinations of variables that span the relevant case space. The choice of which variables to hold constant depends on whether the application involves a military-balance assessment, an arms-control analysis, an evaluation of a new weapons system, operations planning, or something else.

Table 5.1 shows an experimental design for a hypothetical mid-1990s study assessing U.S. and allied capabilities to defeat an invasion. The abstractions, such as political-military scenario, are narrowed to specifics. In this case, the political-military scenario is represented simply by varying the time at which the United States and its ally begin mobilizing and deploying (i.e., at D–10, D–5, D, or D+5). As an-

---

[15]For more details, see Chapter Nine, "Exploratory Analysis and Implications for Modeling."

Table 5.1

Illustrative Experimental Plan: Scenario-Space Analysis of Defense Capability for a Given Name-Level Scenario

Political-military setting	Alliance reacts at C (when deployment begins) = D–10; D–5; D; D+5.
Strategies	Fixed: invasion to specific objectives; defender attempts to halt advance. One measure of outcome is where alliance is able to hold.
Forces	Enemy: 15, 20, 25, 30, or 35 divisions; 10 or 20 tactical fighter wings. Ally: 8 or 12 divisions; no significant tactical air forces. U.S. commitment: 0, 5, or 8 tactical fighter wings; 0 or 5 divisions.
Weapons systems	Ally does or does not have MLRS/DPICM.[a] U.S. does or does not buttress the ally's ground forces with reconnaissance, surveillance, and targeting information. U.S. does or does not have the BAT munition for its MLRS launchers.[b] U.S. aircraft may achieve 1, 2, 5, or 10 kills per day after air defenses are suppressed. U.S. has capability to suppress (either destroy or prevent use of) air defenses in 1, 4, or 8 days.
Environment	Normal weather.
Algorithms and other model assumptions	Attacker's nominal ground force effectiveness (based on equipment) is multiplied by 0.75 or 1 to reflect uncertainties about competence and dedication. Ally's nominal ground force effectiveness is multiplied by 0.5 or 1 to reflect uncertainties about preparation, competence, and dedication.

[a]The MLRS is a multiple-rocket launcher system that can launch a variety of munitions, one of which is abbreviated DPICMS.

[b]The MLRS can also launch the Army's Tactical Missile System (ATACM) missile, which will be able to carry brilliant anti-tank (BAT) munitions.

other example, the design includes cases in which the United States does and does not assist the ally's ground forces by providing timely reconnaissance, surveillance, and targeting information. This can have a factor-of-two effect. Overall, this analysis design would entail running 200,000 cases. This is not traditional sensitivity analysis.

My colleagues and I have used such experimental designs in studies related to NATO's defense capability, the defense of Kuwait and Saudi Arabia, and the defense of Korea. Each design was tailored to the problem at hand. In a mid-90s study, for example, the future Russian threat to Poland was quite hypothetical, and the only issue was to understand how difficult it might be for NATO to defend Poland if

Poland joined NATO. For such a study, we were careful to avoid worst-casing that postulated not only a future malevolent Russia, but also a large and supremely competent Russian army. Instead, we considered somewhat larger-than-expected Russian threats that were reasonably credible for the period under study, but did not ascribe to them advanced, U.S.-like capabilities. We also considered a range of Polish self-defense capabilities, and so on. In contrast, when studying what capabilities might be needed in the more distant future to deal with a regional peer competitor, we drew on intelligence estimates, Defense Science Board studies, and other efforts to define a range of more stressful but *plausible* (i.e., not incredible) threats—just to explore under what circumstances various U.S. capabilities would be especially valuable.

The design of such analyses can benefit from a method called "fault trees" (see Figure 5.6) that was used in a mid-1990s study that examined the defense of Kuwait and Saudi Arabia.[16] The purpose of the

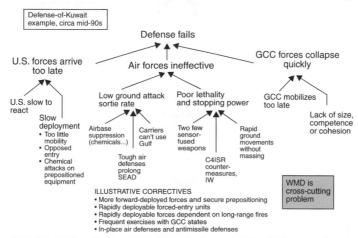

NOTE: GCC = Gulf Cooperation Council; SEAD = suppression of enemy air defenses; WMD = weapons of mass destruction; C4ISR = command, control, communications, computers, intelligence, surveillance, and reconnaissance; IW = information warfare.

Figure 5.6—Fault Tree for Defense of Kuwait and Saudi Arabia

[16]For a more recent study that addresses anti-access issues in some detail, see Paul K. Davis, Jimmie McEver, and Barry Wilson, *Measuring Capabilities for Interdiction: Exploratory Analysis to Inform Adaptive Strategies for the Persian Gulf*, MR-1371, RAND, 2002.

diagram, which can be created in brainstorming sessions, is to organize thinking so as to highlight the factors worth varying in analysis. Figure 5.6 does this by indicating different ways in which the United States and its allies could fail—e.g., U.S. forces could arrive too late, U.S. air forces might not be effective enough to bring about an early halt, or the local Persian Gulf–state (GCC in the figure) forces might collapse quickly because they are outnumbered and outgunned. Lower on the tree are ways these subfailures might occur, and at the bottom of the tree are natural connections to model inputs. For example, in the middle there is reference to tough air defenses prolonging SEAD (suppression of enemy air defenses) operations . In model terms, this means varying the "SEAD time" by considering different values—say, 1, 3, and 8 days.

Figure 5.6 also shows (bottom) a list of illustrative corrective measures for avoiding failed defense. Again, such lists lead to decisions about what model parameters to vary.

Figure 5.7 illustrates a "slice" through the outcome space of a theater-level simulation. Instead of seeing results for one detailed scenario, we see outcomes for a wide range of cases packaged in a way

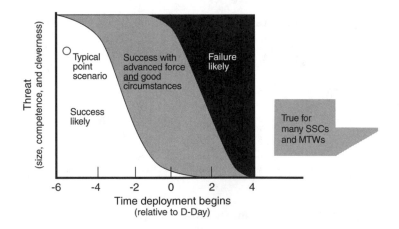

Figure 5.7—One Slice Through Outcome Space, Showing Envelopes of Capability for Defense

that tells an important story. Because we are seeing only a slice through the space, many inputs are held constant, but we do see the effects of varying threat size (y axis) and varying start time for the U.S. force deployment (x axis). Outcome is indicated by white (good), gray (uncertain), and black (likely failure). The principal point is that immediately employable force is particularly critical because fully actionable warning times may be quite short, even though strategic warning usually exists.

This depiction of analysis puts pressure on the observer to favor planning options that would put resources where they are most needed, rather than continuing to work on marginal improvements for cases in which the United States already has good capabilities. In other words, the obvious implication of such a display is to put emphasis on turning more of the shaded region to white and more of the black region to white or shaded. That is, the premium is on *early* capabilities.[17] This is in contrast to spending more money to obtain an even better outcome for scenarios already in the white region.[18]

Choices and Resource Allocation

The last topic of this chapter is how to move from a portfolio construct of overall strategy, capabilities assessment, and assessments of environment shaping and strategic adaptiveness to informing resource allocation choices. After all, capabilities-based planning cannot simply provide a blank check to prepare for any and all possibilities.[19] A portfolio framework helps, but additional methods and tools are needed, some of which already exist.

---

[17]Many options for improving *early* capabilities are discussed in Paul K. Davis, Richard Hillestad, and Natalie Crawford, "Capabilities for Major Theater Wars," in Zalmay Khalilzad and David A. Ochmanek (eds.), *Strategic Appraisal, 1997: Strategy and Defense Planning for the 21st Century*, MR-826-AF, RAND, 1997, and in Gritton et al., *Ground Forces*.

[18]As discussed by E. C. Gritton et al. (*Ground Forces for a Rapidly Employable Joint Task Force: First-Week Capabilities for Short-Warning Conflicts*, MR-1152-OSD/A, RAND, 2000), a generic version of this chart applies well in thinking about capabilities for both MTWs and SSCs (such as the Kosovo conflict).

[19]The blank-check problem is one of those raised by people who are skeptical of capabilities-based planning by "uncertainty hawks" (Carl Connetta and Charles Knight, *Dueling with Uncertainty: The New Logic of American Military Planning*, Project on Defense Alternatives, Commonwealth Institute, Cambridge, MA, 1998).

Figure 5.8 suggests that a portfolio-structured decision-support analysis should have the following inputs: multiple objectives, multiple options, many scenarios and variations (distilled from exploratory analysis), "objective" MOEs (e.g., results of simulations), multiple cost and budget measures, and subjective judgments about effectiveness. Bluntness is appropriate here about a controversial matter: Those who reject subjectivity in methodology have no place in higher-level planning, since the most important decisions are inherently subjective. Further, most allegedly objective measures derive from models with many uncertain assumptions influenced by judgments, or from data that may be a poor proxy for what is needed (e.g., Vietnam-war body-count data). The challenge is not to make things "objective," but to structure subjective judgments so that they are well defined and meaningful as part of an analysis.

Figure 5.9 is a top-level view of a desirable planning structure. The first column lists some of the policy options (these are discrete program-level options, not high-level strategy alternatives); other columns are for their different MOEs. The scorecard approach allows decisionmakers to see the component assessments, rather than forcing them to buy into any particular combining rule for overall utility.

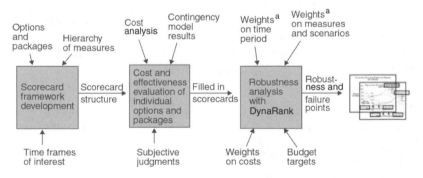

[a]Combining rules may also be nonlinear, as in taking the worst component score rather than using a weighted score.

Figure 5.8—Factors to Be Reflected in Portfolio Analysis

Option (vs. baseline)	Capabilities		Environment Shaping	Strategic Adaptiveness	Net Effect	Cost
	MTWs	SSCs				
+20 B-2s						
+1,500 SFWs						
+Allied package						
+150 UAVs						
CEC acceleration						
+1 CVBG						
More home-porting						
Many others						

NOTE: CEC = cooperative engagement capability; CVBG = carrier battle group; SFW = sensor fused weapon; UAV = unmanned aerial vehicle.

Figure 5.9—Basic High-Level Scorecard Structure

Three of the columns in this illustration correspond to the portfolio construct (capabilities, environment shaping, and strategic adaptiveness). Of these three, capabilities has two subcolumns, one for MTWs and one for SSCs, whereas environment shaping and strategic adaptiveness have none. In another application, they might have several. For other applications, there might be another column for force-management effectiveness (measuring, e.g., effects on operational tempo and ability to recruit and maintain forces). Such matters can easily be changed using a spreadsheet tool such as RAND's DynaRank.[20]

The value in any cell of the scorecard may be generated from subordinate spreadsheets that use simple models or have databases arising from more simulation or empirical work. The top-level assessment of MTW capability, for example, might be the result of numerous subordinate-level calculations for different MTWs with different assumptions, such as those about warning. Those discrete cases, in turn, would have been chosen after broad exploratory analysis in the relevant scenario space revealed the most important

_____

[20]See R. J. Hillestad and Paul K. Davis, *Resource Allocation for the New Defense Strategy: The DynaRank Decision Support System*, MR-996-OSD, RAND, 1998.

factors. Still other subordinate spreadsheets and notes may elaborate on the reasoning used in subjective judgments.

Performing this type of analysis permits us to order policy options by effectiveness, cost, or cost-effectiveness. This requires a rule on how to combine the shaping, capabilities, and preparing-now components. Given such a rule, the analysis determines how to cut costs with minimal impact on effectiveness (left side of graph in Figure 5.10) or how much additional effectiveness can be achieved by buying the entire set of policy options (right side of graph).

Exploratory analysis is as essential to cost-effectiveness analysis as it is to capabilities assessment. It helps to construct different "views," which combine assumptions used in the portfolio assessment with those about the relative weights of shape, respond, and prepare now; the weights of subordinate measures such as MTW and SSC capability; and even assumptions in underlying analyses—e.g., how bad to make the worst-case scenarios. These assumptions should reflect differences in how significant people think about the issues (e.g., the weight of readiness versus modernization).

With such views established, portfolio assessment can determine how the rank orderings of options change from view to view (Figure 5.11). Options that rise to the top for all views are, by definition, ro-

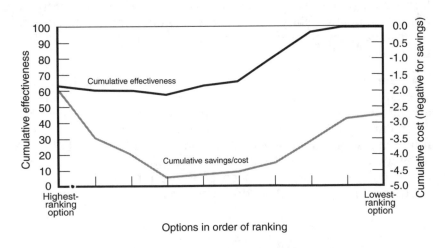

Figure 5.10—DynaRank Outputs in Cost-Effectiveness Analysis

View Emphasizing Warfighting	View Emphasizing Shaping	View Emphasizing Cost
New BRAC	25% fewer F-22s	-1 active division
Double surge sortie rates[a]	New BRAC	-1 CVBG
Allied defense package (helos, ATACMS, advisors)[b]	Allied defense package (helos, ATACMS, advisors)	25% fewer F-22s
Equip F-22s for anti-armor missions	Forward deployed wing	-2 FTWs
Rapid SEAD (HPM)	Mediterranean home-port	New BRAC
Forward deployed wing	Arsenal ship plus allied package	Forward deployed wing

NOTE: BRAC = base realignment and closure; FTW = fighter tactical wing; HPM = particular mechanism involving high-power microwave weapons.
[a] Double surge sortie rates have to do with a program that would permit temporary doubling of air force sortie rates.
[b] An allied defense package consists of specialized units and equipment that would be deployed early to help

Figure 5.11—Exploratory Analysis Showing Rank-Ordered Preferences Under Different Views

bust. Such options usually exist because big organizations have many inefficiencies that make no sense under any sensible view. In the example, all three views recognize the need for another series of base closings (a new base realignment and closure [BRAC] round). As long as the viewpoints are reasonable rather than zealous (e.g., as long as a modernization view still recognizes that shaping is significant), such analysis can help elicit agreement among people who would otherwise argue vociferously in the abstract. For example, in other work of a similar nature, the need to modernize information systems emerged as a consensus view across "stovepipe czars."

This discussion, then, has suggested how it is possible to go from the high concepts of grand strategy down to the nitty-gritty issues of economic choice using one intellectual framework. There is no guarantee that this process of working up and down the ladder of choice will be easy. But it is both feasible and desirable—given strong management, good will, and participation by senior leaders of the defense community.

Chapter Six

PLANNING THE FUTURE MILITARY WORKFORCE

Harry J. Thie

Different questions about military manpower requirements, manpower costs, and trained personnel inventory have emerged as important at various times. For example, just before World War II, the most important question was how to procure a large force immediately. After World War II, the emphasis shifted to managing a large inventory of people with military experience. Now, the emphasis is on recruiting in a highly competitive labor market. Although the broad questions remain fixed, particular aspects of them—recruiting, training, retaining, promoting, compensating, and retiring—receive more or less emphasis at particular times, depending on the nation's military, social, and economic environment. Questions and answers are shaped by many forces, including external events, societal concerns, missions, organization, technology, budget, and demographics. Specific forces at work for the past few years that affect how manpower, personnel, and training questions are asked and answered include a military end strength that is shrinking, threats changing from known to varied, a unitary mission of global conflict shifting to diverse missions within an overall policy of selective and flexible engagement, multiple unit missions replacing single unit missions, variable hierarchies replacing fixed organizational hierarchies, and a joint perspective in operational matters supplanting a service focus.

Planning the workforce is about ensuring that manpower needs are met with trained people at reasonable costs. In past years—as a result of practices from World War II through conscription and the draft-induced volunteerism that ended with Vietnam—planning the workforce was largely about determining its size: forecasting who would stay if they were currently part of it, and calculating how many new entrants would be needed each year. "Manning" the force was the mantra; "aligning" the force by numbers and grades was the goal. Planning the future force is currently shifting away from recruiting and toward overcoming the retention shortfall. Today, workers are scarce throughout the economy—not overall, but among those who are skilled and committed. Emphasis is shifting away from pure recruiting and retention strategies to strategies aimed at developing existing workforces to carry out emerging work. The goal is to develop a cadre of personnel who have the skills needed to meet the requirements of the national military strategy. Human capital accretes; it is smarter to build on what exists than to start anew.

This chapter focuses on manpower, personnel, and training. Manpower deals with the numbers and types of people needed to accomplish missions, personnel involves managing people either directly (e.g., via assignments) or through incentives (e.g., via compensation), and training and development affects knowledge, skills, and behaviors. The chapter also addresses questions about the size of the force; its grade, skill, and experience composition; and the cost and optimal methods to procure, enter, train, develop, assign, advance, compensate, and remove personnel.

To illustrate these issues, an example of how to analyze manpower and personnel policies is worked through. The specific example explores manpower, personnel, and training policy alternatives for achieving an available, qualified, stable, experienced, and motivated future force. Some policy choices can be applied servicewide; others can be applied to military units or to individual soldiers, sailors, airmen, or marines. The example illustrates controllable and uncontrollable variables involved in policy alternatives, relationships among policy choices and desired outcomes, complexity and conflicts among competing objectives, and how undesired outcomes can emerge. Frequently, the complex relationships among variables are not sufficiently considered in the process of choosing policies. As a result, what appears to be a reasonable decision to save budget dollars can have unforeseen future personnel and training consequences.

MILITARY HUMAN CAPITAL

Human capital comprises the skills, knowledge, and abilities of individuals and groups and has value inside an organization and to an organization's customers. The United States has always gained a military advantage by being able to draw on large pools of human capital.

Inside an organization, the correct skills, knowledge, and abilities lead to greater efficiency; outside an organization, they lead to greater effectiveness with customers. In the U.S. military, the combatant commanders take skills, knowledge, and abilities developed by the services and apply them to seek military effectiveness around the globe for diverse missions. The military needs a proper mix of skills and knowledge to gain internal operating efficiencies and to be effective on battlefields, however defined.

Human capital can be thought of as a "stock," much as we think of a stock of materiel. It must be built, maintained, and upgraded; left alone, it deteriorates and becomes obsolete. The next few sections review what the stock of military human capital is now, how it got that way, and what it is likely to undergo in the future, specifically:

- The historical size, source, and composition of the active military

- The present composition, characteristics, and attributes of the active military

- The future effect of the present under likely conditions.

Historical Size, Source, and Composition of the Active Force

Consider, one at a time, various characteristics of the active force: size, enlisted/officer mix, skill mix, and source of manpower. Size largely depends on external events; type of manpower and mix of occupations depend on mission, organization, and technology; and source of manpower—conscription or volunteer—depends on the size of the military relative to the size of the population.

Figure 6.1 shows the size of the active military over a 200-year period. For most of U.S. military history, the number of soldiers, sailors, airmen (after the 1920s), and marines largely depended on external

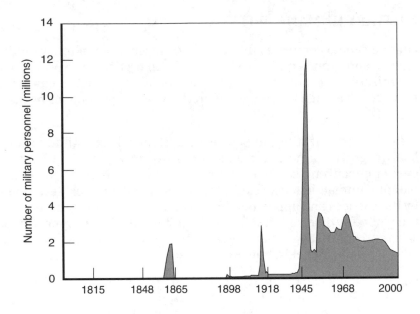

Figure 6.1—Size of Active-Duty Military, 1800–2000

events, with "bumps" in the size of the active-duty force associated with the nation's earliest wars. When the nation needed to increase the size of the military, it enlisted or conscripted recruits from the general population; when the need abated, these recruits were equally quickly separated. For example, the military increased from 28,000 in 1860 to nearly 2,000,000 by 1864; by 1866, the force had decreased to 77,000. The more recent surges in manpower appear as spikes in the data. Notice the sharp spike by 1945, as the nation mobilized for World War II and the military grew to over 12,000,000. Three years later, however, the force had shrunk to about 1,400,000, a level close to today's.

Since World War II, external events have continued to shape the size of the force—the post–World War II drawdown in the late 1940s; the Korean War in the early 1950s, with its own subsequent drawdown; the Berlin crisis, which added manpower in the early 1960s; the Vietnam conflict, with its own era of growth and drawdown. The debacle at Desert One was followed in the 1980s by the Reagan buildup, the fall of the Berlin Wall, Grenada, and Panama. In the 1990s, the breakup of the former Soviet Union, the Persian Gulf conflict, and participation in humanitarian and other nonwar operations were significant. The 1990s witnessed a sharp reduction in the size of the active military.

Figure 6.2 documents the enlisted-to-officer ratio over two centuries. During the 19th century, the ratio was rather flat, with periodic spikes; in times of conflict, the ratio tended to increase as more enlisted personnel than officers were added.

The number of enlisted personnel increased by a factor of almost six, to about 225,000, for the Spanish American War. After that war, the level fell by one-half, to about 115,000, which was a consistent level up until World War I. The "standing" active military began in this era. However, officer strength did not increase as much during this period, which accounts for the high enlisted-to-officer ratio for the first 20 years of the century. Before World War I, brawn still mattered most on the battlefield; coal-fired ships, dismounted infantry, and horse-drawn artillery required proportionally more enlisted personnel. An air force did not exist.

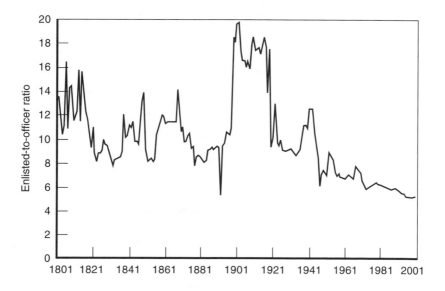

Figure 6.2—Ratio of Enlisted Personnel to Officers, 1800–2000

Since World War I, there has been a significant shift from enlisted to officer manpower. The introduction of the airplane, tank, modern steam ship, and radio shifted work toward more use of brain than brawn. New technologies tend to be "officer heavy" when first introduced, because initially they are complex and require doctrinal and organizational change. Technological innovations also initially require a larger, officer-rich support tail to provide service and supply. Since World War II, the ratio's trend has been downward, with less frequent spikes, as officers have come to represent a larger proportion of a large active military. Moreover, beginning in World War II, the need to coordinate, integrate, and sustain military forces numbering in the millions, rather than tens of thousands, has led to officers being substituted for enlisted personnel, in part to staff increasingly larger and more hierarchical organizations. These broad trends continue to the present.

The organization of work and the composition of the military force are never static; they change with mission, organization, and technology. The columns in Figure 6.3 summarize the changes in the occupational distribution of the enlisted force from 1865 to 2000.[1] The precipitous decline in jobs classified as general military is as striking as the increase in technical occupations and craftsmen.

During the early years of the military, the demand for occupational specialization was small. Almost all soldiers were infantry riflemen, with a few serving in support activities. The Navy was the first to experience the effect of the Industrial Revolution, and the shift from sails to steam was a far-reaching technological change. The Army lagged for several decades, until the World War I mobilization, but its subsequent transformation was dramatic. By 1918, the combat soldier was for the first time in a numerical minority.

Following World War II, several factors dramatically changed the occupational requirements of the services—e.g., the acceleration of weapons and military technology to include the nuclear military, the application of electronics to communications and logistics, and the

---

[1] Harold Wool and Mark Eitelberg are to be thanked for preserving much of this history. See Harold Wool, *The Military Specialist: Skilled Manpower for the Armed Forces*, The Johns Hopkins Press, MD, 1968; Mark J. Eitelberg, *Manpower for Military Occupations*, Human Resources Research Organization, VA, 1988.

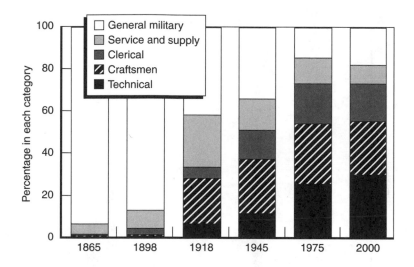

Figure 6.3—Occupational Mix, 1865–2000

emergence of missiles and air defense. Organizational structures changed to take advantage of the new armaments and processes, and there was a noticeable occupational shift from infantry, artillery, and seamanship to technicians. By the early 1980s, technical workers constituted the largest of the five separate groupings, as they do today. As of 2000, 18 percent of enlisted personnel were in a general military specialty, 35 percent were blue collar workers (service and supply workers and craftsmen), and 47 percent were white collar workers (clerical and technical workers).

The United States has used conscription and voluntary enlistment to raise manpower but has never used universal service (although that concept was hotly debated between World War II and Korea).[2] Figure 6.4 shows the active military as a percentage of the population from 1800 to 2000. The horizontal bar running across the lower part of the figure has four lightly shaded segments showing the four times (which together total only 30 out of 210 years of federal history) con-

---

[2]The term *conscription* means that certain rules are used to select a portion of the population for service. *Universal service* (or *universal military training*) means that all militarily qualified citizens must serve.

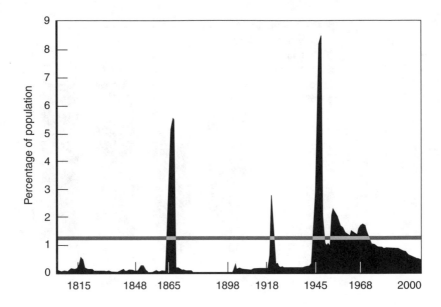

Figure 6.4—Active Military as Percentage of Population, 1800–2000

scription was used. The Civil War and World War I were both periods of wartime conscription; the first use of peacetime conscription was in 1940, shortly before World War II. The nation had a volunteer force from 1946 to 1950, but Korea saw a return to conscription. Vietnam manpower policy was dominated by conscription to meet the needs of the Army; the needs of the other services and the reserve component were met largely with volunteers (though many of these were draft induced).

Figure 6.4 also shows how the size of both the active enlisted force and the population related to the use of conscription (shown in lightly shaded part of bar) over time. The nation has always used conscription when it required an active enlisted force larger than about 1.4 percent of the population, something that has occurred only during the four periods described above. Since 1973, when the last period of conscription ended, the population has grown and the active military has shrunk. Currently, the active force represents about 0.5 percent of the population. Using the 1.4 percent as a rule of thumb means that the need for an active force of above 3.7 million would lead to conscription—a highly unlikely event in the foreseeable future. Also, conditions outlined below will most likely make the rule of thumb obsolete.

The lessons from this history are central to assessments about the future. However, the durability of these lessons cannot help but be affected by more-recent fundamental changes that must be incorporated into any synthesis of the future: the greater use of civilians and reserves, and the greater education and higher aptitude of the enlisted force.

Ongoing Revolutions: Composition, Characteristics, and Attributes

Several "revolutions" now under way have import for the future. The first of these stems from the total force policy, which was about using all the manpower resources of defense. This policy was implemented in 1971 but, as Figure 6.5 suggests, did not see its defining year until 1985—the first year in which Department of Defense (DoD) civilians

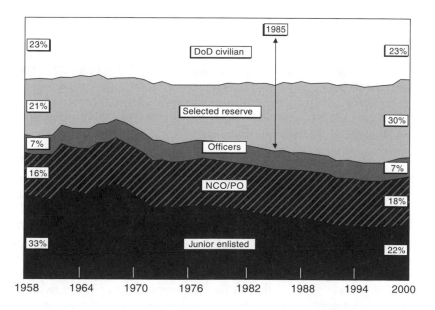

Figure 6.5—Defense Manpower Composition, 1958–2000

and selected reservists (Guard and Reserve), combined, outnumbered the active military. As of 2000, only 47 percent of total defense manpower has been active military. This is a fundamental change that needs to be central to thinking about the future stock of military human capital.

Another fundamental change took place within the active military. From 1958 on, active officers were a consistent 7 percent share of total defense manpower. However, in the enlisted category, noncommissioned officers (NCOs) and petty officers (POs) gained share—up to 18 percent of defense manpower. Civilians, reservists, and NCO/PO have, together, replaced junior enlisted (those in the grades of E1 to E4); the latter have dropped from one-third of all defense manpower in 1958 to about one-fifth in 2000. This follows directly from the changes in the nature of the military: work has become more complex and specialized, and more education and experience are required to do the work successfully.

Another dramatic change is the quality of the force. Questions remain about how best to measure quality and how much quality is really needed or affordable, given the fiscal constraints facing the military. During the post-Korean draft era and the early part of the all-volunteer force, quality (as measured by trainability) tended to be lower than it has been during the modern volunteer era. Once again, 1985 was a defining year. Figure 6.6 shows a revolution that has taken place within the enlisted force: as measured by the percentage in the highest three categories of training aptitude (Armed Forces Qualification Test [AFQT] Categories I–III), the current enlisted force, upon entry, has more ability than it ever had.

During the conscription period from 1952 to 1973, Army entrants typically had the lowest level of aptitude among all service entrants. In the early years of the all-volunteer force, aptitude levels fell significantly as the services, particularly the Army, struggled to recruit. The significant drop in the late 1970s was caused primarily by the entry test having been misnormed: the military was recruiting people who had less training aptitude than it thought. This situation changed significantly in the 1980s, with most services above 90 percent for the decade. And yet the real revolution happened in the 1990s, with all services at 97 percent and above beginning in 1991. Starting in 1985, there were 15 straight years in which at least 90 percent of all recruits

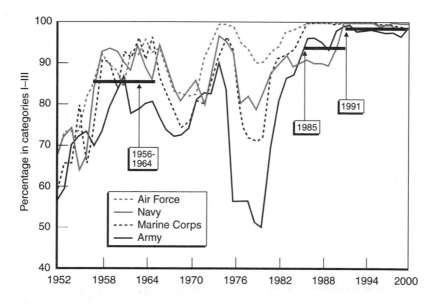

Figure 6.6—Percentage of Enlisted Accessions in Highest Three Training Aptitude Categories, 1952–2000

entering the services were in these three highest aptitude categories. And those who have entered since 1985 will make up over 95 percent of the enlisted force across all services by 2005. Where does this aptitude revolution lead? Virtually all NCOs/POs in the year 2010 (99.5 percent) will have entered since 1985. This highly trainable current and future force bodes well for those who emphasize the learning requirements of the future and the development of military human capital.

Hand in glove with the aptitude revolution is an education revolution. Officers have always been mostly college graduates, but NCOs and POs are now becoming more educated as well. In recent years, officer duties have devolved to NCOs and POs. Many types of work, from administration and paperwork to missile launching, have become sergeant and PO work—largely because of technology and a faster operational tempo. Aspects of this devolution of duties can also be seen in changing concepts of development. Training, a skill-based concept that results in an immediate ability to do certain tasks more proficiently, is slowly giving way to education as a knowledge-based enlisted force increasingly becomes a necessity for successful performance.

The enlisted force is more highly educated than in any previous era. Figure 6.7 shows the proportion of grades E6 to E9 with at least some college education for three different time periods. Fiscal year 1972 represents a force that, while not necessarily conscripted, was shaped by conscription; fiscal year 1981 represents a force shaped additionally by the early years of the volunteer force; fiscal year 1996 represents a force almost exclusively the product of the volunteer era. The figure shows that, generally, the higher the grade, the higher the educational achievement. This figure and the earlier ones show that the proportion of college graduates grew over time as well as across all grades and occupations.

Figure 6.8 shows the data by occupation (aggregated for grades E6 to E9). Even in the general military occupational group, which is the group associated with Army and Marine Corps combat skills, the proportion with at least some college is much greater than was generally believed to be achievable in the era of a conscription-shaped enlisted force.

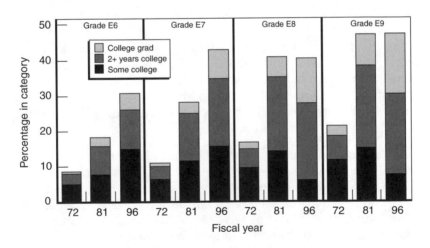

Figure 6.7—Educational Achievement for Grades E6 to E9, Selected Years

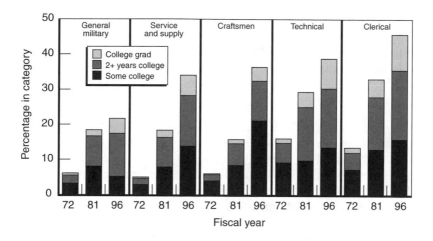

Figure 6.8—Educational Achievement for Grades E6 to E9, by Occupation, Selected Years

Looking to the Future

The near future is pretty much determined by the present and by recent history. Consider the most likely near future, in which the world of 2010 reflects today's trends—e.g., (1) the large U.S. military shrinking in size, (2) known threats becoming varied threats, (3) the unitary mission of global conflict becoming diverse missions within an overall policy of selective and flexible engagement, (4) single missions for units becoming multiple missions for units, (5) variable hierarchies replacing the fixed organizational hierarchies, (6) advanced weapons becoming integrated systems and processes, and (7) a service focus in operational matters continuing to be replaced by a joint perspective. Some speculations about more-radical excursions based on size, organization, and technology follow.

In the broadest sense, the military does what the nation asks: all missions, against all enemies, as the oath requires. Yet not all missions have to be done full time in large numbers or in all types of functional units—a distinction that differentiates active from reserve. The future military will continue to wear uniforms and, in the words of a Navy admiral, "get shot at." This differentiates the military from the defense civilian and contractor workforce (even given that the first casualty in Somalia was a defense civilian and the first in

Afghanistan was a CIA operative). The reduction of active military manpower as a proportion of all defense manpower is likely to continue. More reductions of direct defense manpower will occur as more contractors are used on and off the battlefield. Eventually, the manpower needs of the active military will be about one-third of a percent of the population, a level easily met by volunteers entering military service.

Officers "lead" in the broad sense of leadership. They set direction, align forces, and empower people. This differentiates the officer corps from the NCO and PO corps, which has evolved in its own right to be the force's managers and technicians. The experience level of the enlisted force will continue to grow as NCOs and POs continue to displace junior enlisted. Moreover, junior officers who have been managers and technicians more than leaders will begin to be supplanted by educated and experienced enlisted technicians and managers. So far, with enlisted personnel acquiring more status, their work appears to be taking on more of the hallmarks of a profession and fewer of the hallmarks of a trade. As the edge in human capital enjoyed by junior officers over senior enlisted shrinks, many issues related to the scope of the latter's responsibility, job design, status, and compensation become salient.

Two enduring characteristics—cognitive ability and conscientiousness—explain most variability in job performance. These characteristics are largely inherent in the individual and not easily changed later in life. People have them in varying levels when they enter organizations; one cannot develop, train, or educate for these characteristics except at the margin. Therefore, people have to be selected for them. The military has had and will continue to have a "select the best" strategy, expecting all entrants to be able to rise to the highest levels of responsibility. The military strives for a homogeneous entering group of high ability. As better measures of such characteristics as conscientiousness are found, the military will select for them as well. Neither all militaries around the world nor all organizations can afford such a strategy, however; instead, they select a mix of people with varied levels of these characteristics. But to get people with high levels of these characteristics, one must have appropriate selection and compensation instruments in place.

Occupational classifications derive from what work the military will be asked to do (its missions), how it structures itself to do the work (its organization), and the systems and processes it uses to accomplish the work (its technology). The competencies for military human capital change over time. Doctrine and theory for military and naval science and paradigms for leadership and management are not constants. Miniaturization, digitization, and other advances change needed military expertise. Tanks replace horses; steam replaces sail; turbines replace pistons. Is technology part of military knowledge and expertise? Certainly. Technology—how things are done—has always been part of the core knowledge of the military profession. Reskilling and upskilling will enter the lexicon of force management. The highly trainable and experienced enlisted force will be developed and redeveloped to ensure it possesses attributes critical to mission performance.

The military is likely to become smaller, more experienced, and more highly graded, with the percentage classified as general military or line category reaching a new low. These grade and skill percentages will vary by service, as they do now, but the differences will become more pronounced toward 2010. The military might evolve in a different way, however, because of varying size (smaller or larger), organizational change (outsourcing, streamlining, downgrading), and technology (user friendly or not user friendly). Based on the direction of change, the military can shrink or grow as well as compose itself of various skills and grades.

Whether military human capital will change depends on how change is defined. If it is defined as a succession of doctrinal and organizational changes in a constant-sized force, the manpower and personnel change will largely be one of composition, grade, and skill, with some manpower reductions because of productivity gains. However, if manpower is reduced to pay for new systems, what remains will be older, have a different mix of skills, and be forced to rely more on reserves and civilians. Either way, trends in manpower categorization, experience, and skill mix will be accelerated, but there will be no fundamentally new prospect.

In sum, this is what the future holds: the stock of military human capital needed will change, but in discernible directions. The active military will remain robust in size, but proportionally more defense will be provided by civilians, contractors, and selected reservists. NCOs and POs will continue to displace junior enlisted and begin to supplant junior officers. Those who enter the active military will be selected on the basis of certain enduring characteristics, such as cognitive ability and conscientiousness, for which a good measure is needed. They will be well-trained and educated. Specific knowledge and other developed attributes will change as the core of professional knowledge and experience that all military personnel must have changes with mission, organization, and technology. It is likely that career management and compensation practices will need to change. It is not likely that a revolution in military affairs (RMA), reasonably defined, will alter such trends much, but it may accelerate them.

MANPOWER AND PERSONNEL ANALYSIS

To answer questions about planning a future force, one must decide on objectives, criteria, alternatives, and measures. Models and analytic techniques can be used to relate these four elements in useful ways and to draw out consequences and tradeoffs.

The Process

The analytic process begins with the front end (more art than science), in which the problem is identified and structured. The problem first has to be stated appropriately and clearly, at which point it becomes critical that the objectives be articulated so as to spawn more-precise criteria. What is the desired outcome? Almost all objectives deal with some variant of effectiveness or cost. One part of the front end entails devising a broad, creative range of alternative policies that can be evaluated. A large part deals with sorting out ends from means from alternatives from constraints. Constantly asking "why" helps to construct a hierarchy that allows for analysis to enter a structured picture.

In the second part of the process, the analytic middle (muddle?), science replaces the art of the front end. The science does not have to be complex; logic is often sufficient and may be adequately done on the back of an envelope. This analytic middle sorts through consequences: How well do alternatives meet criteria? Models, methods, data, and qualitative assessments are brought to bear. Once the consequences of alternatives are known, tradeoffs can be made. Rarely are all objectives satisfied by one dominant alternative, so tough choices and compromises must be made when all objectives cannot be met at once.

Which objectives are more important than others? An alternative that provides more effectiveness at a lower cost is a no-brainer, but such an alternative is seldom found in the real world; and the reverse, less effectiveness at a higher cost, is seldom chosen. Alternatives that offer more effectiveness at a higher cost or a lower cost for less effectiveness or other possible combinations require hard thinking aimed at determining which alternative appears to best meet the ultimate objectives. Then, one must deal with uncertainties. What could change the assessed consequences? Specific cases can be examined or certain parameters can be judgmentally varied to see how the consequences and thus the conclusion might change. Last, what is the appetite for risk? Is it better to do nothing if there is no stomach to see the consequences through?

At this point, the back end of the analytic process is reached. Analyzing stops, and decisions or recommendations are made—which includes linking them to other problems and solutions in useful ways. Art and science give way to decision and judgment, a process that in the public sector, is not simply about selecting the "best" alternative for maximizing or minimizing criteria related to objectives, but involves consensus. What can the disparate stakeholders within DoD or across the executive and legislative branches agree to? What can be implemented?

The various parts of the manpower community go about their work differently. The personnel community, which seeks the best answer for the long term, studies the long-range, ideal effects of policy or changed parameters. This steady-state analysis applies constant rates and planning factors to see what happens in the steady state when all the transitional effects are settled out. Of course, no state is steady; transitional effects form a constant stream. But steady-state thinking is the best analytic tool for policy optimization.

The program and budget community, however, is more interested in the dynamic effects of a changed policy. What happens over a particular time horizon as rates and factors change successively? Not surprisingly, the five- to seven-year horizon of the program objectives memorandum (POM) process is frequently used. Models show the effects of policy choice projected out for about five years, but not beyond (the famous "straight-lined" from the fifth year). Program and budget analysts are more concerned with the immediate resource consequences than with long-term effectiveness issues.

And as for the managerial community, it is more interested in execution: How do we get to a chosen new policy over a period of time? What decisions have to be implemented and when if we are to gain the benefits and avoid the costs? Unfortunately, deciding on a course and implementing it are not one and the same, so tough choices often are not followed through on. The reality is a bewildering agglomeration of half-completed past policy implementations that confound not only the managerial community, but the personnel and program and budget communities as well. Personnel policy tends to accrete over time in marginal ways.

Choosing Among Alternatives, an Example

To illustrate the process just described, consider the following questions: How do specific changes in deployment, or personnel-related, policies and procedures affect

- Force readiness, particularly the availability of individuals and units for deployment?
- The level and distribution of operations tempo and its related stresses on units and individuals?
- Related outcomes, such as geographic stability, unit cohesion, and career-long job satisfaction and quality of life?
- Retention?
- Cost?

Concentrating on these questions helps in identifying objectives and alternatives—part of the front end of the process. Objectives—outcomes desired by the organization—are of extreme importance in decisionmaking. Although they are usually stated broadly, are usually poorly quantified, and may conflict with each other, their purpose is to establish what the decisionmakers consider important. Here, seven objectives have been outlined: high readiness, high geographic stability, less force stress, high unit cohesion, acceptable quality of life and career satisfaction, high retention, and lower cost. Each objective has an associated directional modifier that is useful for clarifying what the decisionmaker means.

The next part of the front end involves making these broad objectives more precise—turning them into variables that can be measured on some scale or metric. This is part of the art of analysis. One or more variables are chosen to represent each entire objective statement. Some objectives (e.g., cost) present many measurement choices; others (e.g., quality of life) seem inherently unmeasurable given the state of the art. One needs to be clear about why the specific variables and measures were chosen, because they may be assailed by those who disagree with the conclusions reached.

Table 6.1 shows the variables and measures for the seven objectives in our example.

Table 6.1

Variables and Measures for Objectives

Objective	Variables and Measures
High readiness	SORTS P-status distribution; deployable-unit distribution
High geographic stability	Time-on-station distribution
Less force stress	DEPTEMPO distribution; time away from bunk
High unit cohesion	Cross-leveling amount
Acceptable quality of life/career satisfaction	Career deployment/assignment patterns
High retention	Continuation rates
Lower cost	Cost

NOTE: SORTS (Status of Readiness and Training System) uses a numerical assessment of personnel fill to get a calculated readiness status (e.g., P1, P2). DEPTEMPO is the tempo of deployments for a unit; time away from bunk is the tempo for individuals. (DoD is in the process of making these two terms more precise.) For unit deployments, units are frequently "cross-leveled" as deployable individuals are reassigned from nondeploying to deploying units.

The last piece of the front end of the analysis entails defining policy alternatives—i.e., courses of action that, if chosen, could change the measure of an objective. For this example, policy alternatives that might apply to an entire military service were outlined, as were policy alternatives that might apply to certain units within a service and to individuals. Tables 6.2, 6.3, and 6.4 show these three alternatives sets, respectively.

Table 6.2

Policy Alternatives Applicable to an Entire Military Service

Policy	Alternative
National military strategy	More deployments; fewer deployments
Reserve component use	More; less
Active/reserve force mix	More active; more reserve
Unit basing	More dispersed; more concentrated
Proportion of operating forces (TOE/TDA[a]; ship/shore)	Higher; lower
Size of units	Larger formations; smaller formations
End strength	Higher; lower
Grade plan	Richer; leaner

[a]TOE = table of equipment; TDA = table of distribution and allowance.

Table 6.3

Policy Alternatives Applicable to Units Within a Service

Policy	Alternative
Resourcing level	Higher authorizations; lower authorizations
Manning priority	Higher; lower
Tour lengths	48 months; 36 months; 24 months; 12 months
Assignment tenure limit	6 years; 5 years; 4 years
On-deck unit system	Yes; no
P-status deployment standard	P-1; P-2; P-3
Cohesion criteria	No cross-leveling; various percentages of allowed cross-leveling
Cross-leveling rules	Local units only; same theater units only; no restrictions
Deployment priority rules	Longest time since previous unit deployment; highest proportion of deployable people; highest readiness status

Table 6.4

Policy Alternatives Applicable to Individuals

Policy	Alternative
Tour extension incentives	Strong; weak; none
Cumulative annual deployment limits	210 days; 180 days; 150 days; 120 days; 90 days
Administrative deferments	More; less
Deployment deferments	6 months; 12 months; 18 months; 24 months
Reenlistment goals	Higher; lower
Retention control points (high year of tenure)	Earlier; later
Selective reenlistment bonuses	More; fewer

Once the front end portion of the process is finished, the middle portion begins, the purpose being to determine the consequences associated with the different alternatives. In what ways do the measures or criteria change as choices are made? Here, more formal techniques are introduced. Each analysis leads to its own methods for determining the consequences of the alternatives vis-à-vis the objective criteria. Because this sample analysis was complex, a formal simulation model was built to capture all of the complicated relationships among the objectives, alternatives, measures, and policies. Any given model will only capture a subset of all the possible relationships; doing more would take more time and effort.

Figure 6.9 shows the relationships that were modeled. The figure is highly detailed, to be sure, the point being to show that many variables affect other variables. Moreover, not all possible relationships are captured. The model is complex enough to represent all those that were captured.[3]

Like all good models, this one permits sensitivity analysis—i.e., it explicitly shows how the objectives would be affected if different alternatives were chosen. In this sample case, analysis ceased at this point, and a good decision dealing with uncertainties, risks, and implementation issues followed. This case dealt with broad objectives and large-scale alternatives; much of manpower and personnel

---

[3]This model was done in conjunction with a RAND colleague, Al Robbert.

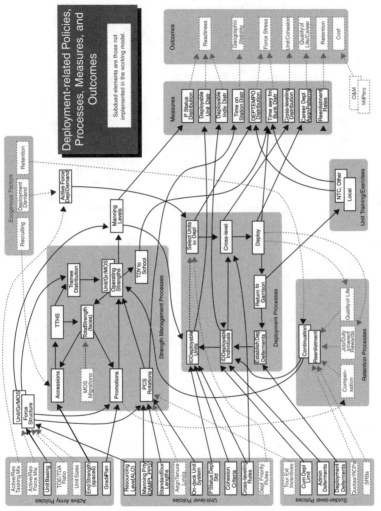

Figure 6.9—Model of Complex Relationships

analysis is narrower and more precise. While trends and broad forces do not change quickly, there is much to be learned by analyzing their impacts when seeking improvements to manpower, personnel, and training policies and processes that will increase military effectiveness and reduce costs.

Chapter Seven

THE SOLDIER OF THE 21st CENTURY

James R. Hosek

Versatility and leadership top the requirements list for the soldier of the 21st century. The future may or may not threaten a major war, but it requires the ability to fight and win one, as well as to engage in a wide variety of smaller conflicts and other operations, such as peacekeeping. The range of demands and flow of new technology call for personnel who can learn rapidly, reach high levels of competence, adapt in the face of uncertainty, and apply a variety of skills in difficult circumstances.

To obtain a versatile, well-led force, there must be a systematic approach for identifying the factors that determine versatility and leadership and for creating the organizational structures and personnel policies needed to support the desired levels of versatility and leadership. While versatility and leadership are inherently the product of policy decisions, it cannot be taken for granted that the requirements for versatility and leadership are well understood, or that the best policies for achieving those requirements are known, or that the policies are actually in place. More to the point, it is probably far more important that adaptable processes be built than that there be a single vision of future requirements and a commitment to that vision.

Adaptability's value stems from uncertainty about the scope of contingencies, the nature of security threats, the speed and range of advances in technology, and the private-sector demand for the kind of members the military would like to recruit and retain. Dynamic workforce planning models therefore have advantages over rigid, input-output planning models. Dynamic models can select policies that are robust to uncertainty and find the least-cost path to changed future requirements. The modeling technology exists, so the feasibility of using such models is not in question. What is important is how to assure that the model's data requirements are satisfied and that the model's structure supports the analysis of policy alternatives.

This chapter is not about a particular model, but about factors relevant to developing and supporting the use of dynamic models in defense manpower planning. The discussion looks to the future, to the past, and to theory. Different kinds of future forces that may affect the quantity and quality of personnel needed are described, and the experience of the all-volunteer force that bears on the services' ability to attract and manage high-quality personnel is reviewed. Among other things, this history includes a Department of Defense (DoD) willingness to conduct controlled experiments and field studies to learn about the cost and effectiveness of policy alternatives. Finally, the chapter discusses theoretical concepts underlying the structure of military compensation, a key policy instrument for meeting future manning requirements. Although the common table of basic pay has been the core of military compensation, larger pay differences across occupations and by experience will likely be needed if career lengths are to differ by occupation.

VISIONS OF THE FUTURE

To imagine the range of future needs, consider visions of future forces: the cyber soldier, the information warrior, peace operations, the rapid response force, low-manning vessels, and evolutionary change. These visions will, no doubt, change over time; they are set out here only to emphasize the diversity of requirements they imply. These visions typify a range of responses to the emergence of the new world order and the advance of technology, with its application to better defense. They are not, however, alternatives to one another: they could all occur.

Cyber Soldier

The cyber soldier belongs to a small unit whose mission is to penetrate enemy territory, engage in surveillance and reconnaissance with the assistance of advanced sensors, and call in remote-fire precision guided weapons. Cyber soldiers will need to be of high quality. The introduction of cyber units will involve organizational changes to assure smooth connections from the unit to the command and control (C2) centers. The cyber soldier concept relies on foreseeable advances in sensor technology, integrated and secure information networks, and computer software capable of tracking large numbers of targets, assigning weapons to them, and firing via remote command. The cyber soldier is entrusted with the decision to fire costly (e.g., million-dollar) weapons and can operate in a stealthy fashion, avoiding detection and capture. The technologies that enable the cyber soldier mean that greater lethality can be achieved with fewer people, with fewer casualties, and, perhaps, at a lower cost (depending on weapon cost and accuracy, as well as how correct the decisions to fire those weapons are). Thus, the cyber soldier concept could reduce the need for large ground forces. The small groups of special operators that fanned out in Afghanistan during 2001–2002, serving as spotters for bringing distant weapons to fire on Taliban and Al Qaeda opponents, were a foretaste of the cyber soldier.

Information Warrior

Enemies will employ asymmetric tactics to harm the United States. In this particular vision, a small number of hackers and terrorists may seek to disrupt information systems in the armed forces and national economy. To counteract this threat, the armed forces have a cadre of highly capable personnel, including contractors, that help develop secure systems, restore systems that have been attacked, trace the source of the attack, and assist in apprehending the attackers. In addition, there is an information warfare capability. Information warriors are up-to-date experts in hardware and software and develop a detailed knowledge of the vulnerabilities of an adversary's information infrastructure. Like pilots and doctors, information warriors have excellent private-sector opportunities (e.g., in safeguarding corporate information systems). Information warriors may be organized into their own units, much as the Air Force has organizations for space systems and missiles.

Peace Operations

Peace operations include disaster relief, humanitarian operations, peacemaking, peacekeeping, nation building, border patrol, and relationship strengthening (such as joint planning exercises and the assignment of U.S. personnel as advisors). With frequent peace operations comes the requirement for a force structure capable of handling several different small-to-medium contingencies at any given time. For some peace operations—such as disaster relief, humanitarian operations, nation building, and peacekeeping—having personnel on the ground is essential, and they need few skills or aptitudes beyond what they have today. But for other peace operations—such as those involving guerillas, urban terrain, large and well-armed adversaries, or chemical, biological, or nuclear weapons (i.e., weapons of mass destruction, or WMD)—special skills and tactics are required. Depending on the variety of peacekeeping operations needed, the services may form specialized units of personnel who are qualified in multiple skills and who take part in unit training for a diversity of threats.

Rapid Response Force

Because a rapid response can reduce the scope, risk, and cost of a major contingency, each service maintains units to deploy on short notice. These units may not require a different quantity or quality of personnel, but they will have undergone unit training and combined-unit training to maintain a high level of readiness. Current training focuses on the possibility of deployment to the next likely trouble spot. To support the rapid response units, logistics are flexible and lean. Supply and repair capabilities are efficient, with no long waits for parts or supplies crucial to the mission. The Air Force, for example, has reorganized as an expeditionary force, a change consistent with reduced overseas basing and an awareness that contingencies can occur anywhere.

Low-Manning Vessels

Labor-saving technologies will allow the services to fulfill their traditional roles and missions with fewer personnel. The Navy builds highly automated vessels that need few crew members. For every combat position eliminated, two to three fewer combat support crew members are needed. And this reduction, in turn, means a reduced number of support personnel are needed for laundry, mess, and the like. Redundant operating systems help sustain the vessel in case of attack or system failure. Crew members need to know how to operate at least one system and, often, two—e.g., automated weapons, navigation, or propulsion, and responses to system breakdown or damage from attack. Even advances in paint make a difference in the number of crew members required on older Navy ships: highly durable paints reduce the number of Gen Dets (general detail personnel) required to maintain a vessel. Other services also benefit from new technology. For instance, an increased reliance on drone vehicles for reconnaissance reduces the demand for pilots, crews, maintenance, and repairs (although it may also increase the demand for intelligence and communications personnel).

Evolutionary Change

Some variant of major theater war (MTW) is always in the set of future planning scenarios. The prospect of MTW points toward evolutionary change, a slow, formidable process that includes force modernization, the incorporation of maturing technologies, and the gradual impact of these two elements on organization and doctrine. Preparedness for major war remains the benchmark for judging military readiness. Not surprisingly, the rationale for many military assets—tanks, helicopters, aircraft, missiles, ships, submarines, artillery, military hospitals, etc.—comes from major war. Peace operations are lesser-included cases that can be handled by a subset of the forces and resources. Moreover, change can be hard to accelerate. The acquisition life cycle of new weapons is about two decades. As force modernization occurs, old equipment dating from the 1950s through the 1970s will be replaced by versions of equipment whose development began in the 1980s and 1990s. Thus, procurement over the next five to 10 years is unlikely to bring surprises in the form of wholly new equipment or systems not included in today's acquisition outlook.

Modernization involves a host of changes: digitization of the battlefield; the use of advanced information systems in logistics, administration, and medical care; and improved precision guided munitions

(PGMs), increases in stealth, improved body armor for personnel, and so forth. To many observers, the combination of advances in information technology, engineered materials, and sensors provides the foundation not just for an evolution but for a revolution in military affairs (RMA). Yet change is usually gradual and tends to lead to functions that are similar but have greater capability, a fact that has two implications for defense manpower. First, if today's personnel satisfy current military requirements, they should be adequate in the near term—i.e., the same knowledge, skills, and aptitudes are likely to be appropriate. Second, there should be enough time to change the kind of personnel or their preparation as needed. This does not necessarily mean, however, that today's personnel management and compensation policies are well suited to support such change, or even that the policies are sufficient to maintain the status quo.

Table 7.1 summarizes the different visions with respect to the need to change the organizational structure, the use of new technology, the impact on total manpower requirements, and the demand for high-quality personnel. Most changes will probably be evolutionary, yet change in technology, doctrine, or strategy might lead to radical change in manpower requirements for specific force elements.

Table 7.1

Comparing the Six Visions of the Future in Terms of Needed Changes

Vision	Organizational Change	New Technology	Manning Requirements	Personnel Quality
Cyber soldier	Yes	Yes	Decrease	High
Information warrior	Maybe	Yes	Increase	High
Peace operations	Maybe	Maybe	Same	Same
Rapid response	Yes	Maybe	Same	Same
Low-manning vessels	No	Yes	Decrease	High
Evolutionary change	Gradual	Gradual	Gradual	Gradual

EXPERIENCE OF THE ALL-VOLUNTEER FORCE

If versatility and leadership are needed in all visions of future forces, it is worthwhile to reflect on what these concepts mean. To a great extent, both depend on the quality of service members recruited and retained. In the years of the all-volunteer force, a great deal has been learned about why quality is important to military capability and how much policy action (or inaction) can influence the quality of the force. That knowledge is the background for the analysis and choices ahead.

Versatility and Leadership

Versatility is the ability to engage in multiple activities. At the individual level, training, experience, and aptitude contribute to versatility. Training provides the knowledge and skills required for certain tasks performed on duty assignments; military training typically consists of basic training, advanced individual training, and on-the-job training. Depending on the specialty's complexity, an enlisted person progresses from entry to intermediate to high skill level. The Air Force describes these as skill levels 3, 5, and 7, for example, which are analogous to apprentice, journeyman, and master levels in trade unions. As their experience increases, personnel work on different tasks, equipment, and missions in a wider range of activities, and their versatility increases. Aptitudes measured include, among others, verbal, quantitative, spatial, mechanical, and coding speed (how fast a person can assimilate new information). The U.S. military measures aptitudes with the Armed Services Vocational Aptitude Battery (ASVAB); it summarizes verbal and quantitative aptitudes in the Armed Forces Qualification Test (AFQT) score.

The concept of versatility goes beyond the individual service member. Peace operations in the 1990s often did not require units to perform missions different from their MTW missions; peace missions were mostly a subset of war missions. But the organizational versatility to handle frequent peace operations and yet maintain readiness for MTW was limited. Peace operations created organizational stress because they interfered with training and exercise programs, affected the planned rotation of personnel overseas, and sometimes reduced the quantity and quality of equipment available to nondeploying units. Moreover, there was no budgetary process for prompt funding of the services' added cost of peace operations, and the objective, scope, and duration of certain peace operations were uncertain. Over time, however, peace operations may have induced the services to become more agile and to handle such operations at lower budgetary cost and with less impact on readiness. In addition, peace operations may have provided information about the effectiveness and vulnerability of current doctrine, equipment, and training.

Because versatility depends on training, experience, and aptitude, it comes at a cost. Senior, experienced personnel cost more than junior personnel, and high-aptitude personnel, who often have better private-sector opportunities, will be bid away unless a military career offers sufficient opportunity for advancement and compensation. Thus, when the services make decisions about desired recruit quality and force experience mix, they are implicitly making decisions about versatility and cost.

Leadership skills, like other human capital, can be strengthened through training, education, and experience. In addition to having specific knowledge about an area (e.g., gunnery, logistics, medicine, intelligence), a leader should be able to identify key objectives, allocate resources efficiently toward them, foster unit cohesion, and motivate personnel to perform at high levels.[1] Allocative efficiency involves balancing marginal gains against marginal costs in a decisionmaking environment that is typically dynamic and uncertain. Poor allocative efficiency implies the leader is not making the most of versatile personnel.

Unit cohesion concerns how well a team functions as a result of its members' knowledge of one another's capabilities and commitment to a common objective.[2] In one view of combat leadership, individual performance follows from cohesion because "unit members are bonded together in their commitment to each other, the unit, and its purposes."[3] But this view does not suggest a mechanism for how cohesion is created. According to another view, cohesion will increase if

---

[1]According to William Darryl Henderson (*Cohesion: The Human Element in Combat*, National Defense University Press, Washington, DC, 1985, p. 11): "The leader must transmit organizational goals or objectives effectively from the chain of command to the small, cohesive group. Then he must lead the unit in achieving these objectives through his personal influence and technical expertise. The leader must also maintain unit cohesion by ensuring continuous organizational support and by the detection and correction of deviance from group norms. Finally, the leader assists in making or maintaining an ideologically-sound soldier by setting an example, by teaching, and by indoctrinating."

[2]This describes *task* cohesion, not *social* cohesion.

[3]Henderson, *Cohesion*, p. 23.

a leader can improve performance: "While cohesiveness may indeed lead the group to perform better, the tendency for the group to experience greater cohesiveness after successful performance may be even stronger."[4] A leader can presumably improve performance, and hence cohesion, by improved allocation of unit resources under uncertainty.

Attracting Quality Personnel

The volunteer force is premised on the ability to set military pay high enough that the supply of volunteers equals manpower requirements. Before the volunteer force, conscripts were paid below their average market wage. Moreover, even if their pay had equaled the average wage, the involuntary nature of conscription meant that a sizable portion of entering personnel would still have paid a "conscription tax"—i.e., the difference between the wage required to induce the person to enlist voluntarily and the military wage. Individuals with a high market wage or a low taste for the military had high reservation wages and therefore paid a high conscription tax. It is not surprising that first-term attrition was high among conscripts. Countries that still rely on conscription, such as Germany, Italy, and Russia, set a short mandatory term of service (a year or less). They thus have high turnover and low experience in their junior force.

The shift to an all-volunteer force in 1973 required a large increase in entry pay. Monthly pay for an enlistee rose from $144 in January 1971 to $288 in January 1972 and $307 in October 1972. As a result, the cost of first-term personnel, who made up about 50 percent of enlisted personnel, became a greater factor for a service to consider in determining personnel force size and experience mix.

The volunteer force began successfully but then faltered, in the late 1970s, as military pay fell relative to private-sector pay. This caused the services difficulty in meeting their recruiting and retention targets, and the quality of recruits declined, particularly in the Army. Lower-quality recruits had higher attrition rates and were less likely to complete training successfully. Given these circumstances, the

---

[4]Brian Mullen and Carolyn Copper, "The Relation Between Group Cohesiveness and Performance: An Integration," *Psychological Bulletin*, Vol. 115, 1994, pp. 210–227.

services and the Office of the Secretary of Defense (OSD) initiated research on whether higher-quality soldiers were cost-effective. The research included controlled experiments on enlistment incentives for the active and reserve forces, controlled trials that measured the effect of quality and experience on the performance of mission-essential tasks, and specially designed surveys and field studies related to the relationship between experience and productivity. The tools and methods used in this research were necessary to obtain solid, unbiased estimates of policy alternatives that in most cases had not been tried before and thus could not be studied using historical data.

While the research was under way, Congress boosted military pay by 26 percent between 1980 and 1982, increased recruiting resources, and raised individual enlistment incentives. The Army and Navy introduced their "college funds" to supplement the basic educational benefits available from the GI Bill, and all services made greater use of enlistment bonuses. Reenlistment bonuses were increased in size and offered to more specialties. In addition, the 1982 recession spurred both recruiting and retention. Within a few years, these factors produced large improvements in recruit quality and the retention of experienced personnel.

Table 7.2 summarizes the results of selected analyses of personnel productivity.[5] Of the four indicators of personnel productivity shown—education, AFQT score, experience, and unit stability—the first two are the foremost measures of enlisted personnel quality. Recruits with a high school degree who also score in the upper half of the AFQT score distribution are classified as high quality. Unit stability means a slow turnover of personnel in the unit and hence more experience together on average. Whereas education, AFQT score, and experience are person-specific measures of quality, unit stability is not.

---

[5]The table is mainly based on articles cited in John T. Warner and Beth J. Asch, "The Economics of Military Manpower," in Keith Hartley and Todd Sandler (eds.), *Handbook of Defense Economics, Volume I*, Elsevier, New York, 1995, pp. 368-373, although several other articles were also used.

Table 7.2

Factors Affecting Personnel Productivity

Output Measure	Education	AFQT Score	Individual Experience	Unit Stability
Sorties			√ (a)	
Mission capable rates	√ (b)	√ (b)	√ (a)	√ (c,d)
Maintenance downtime	√ (b,c)	√ (b,c)	√ (b)	
Multitasking			√ (e)	
Supervisor ratings			√ (f)	
Job performance tests	√ (g)	√ (g,h,i,j)	√ (k)	√ (h)

[a]A. Marcus, *Personnel Substitution and Naval Aviation Readiness*, P-3631, Center for Naval Analyses, 1982.

[b]Laura I. Junor and Jessica S. Oi, *A New Approach to Modeling Ship Readiness*, CRM 95-239, Center for Naval Analyses, 1996.

[c]S. Horowitz and A. Sherman, "A Direct Measure of the Relationship Between Human Capital and Productivity," *Journal of Human Resources*, Vol. 15, pp. 67–76, 1980.

[d]Russell W. Beland and Aline D. Quester, "The Effects of Manning and Crew Stability on the Material Condition of Ships," *Interfaces*, Vol. 21, pp. 111–120, 1991.

[e]Glenn Gotz and Richard E. Stanton, *Modeling the Contribution of Maintenance Manpower to Readiness and Sustainability*, R-3200-FMP, RAND, 1986.

[f]Mark J. Albrecht, *Labor Substitution in the Military Environment: Implications for Enlisted Force Management*, R-2330-MRAL, RAND, 1979.

[g]Thomas V. Daula and D. Alton Smith, "Are High Quality Personnel Cost Effective? The Role of Equipment Costs," *Social Science Quarterly*, Vol. 73, pp. 266–275, 1992.

[h]Barry L. Scribner, D. Alton Smith, Robert H. Baldwin, and Robert Phillips, "Are Smarter Tankers Better? AFQT and Military Productivity," *Armed Forces and Society*, Vol. 12, pp. 193–206, 1986.

[i]Bruce Orvis, Michael Childress, and J. Michael Polich, *The Effect of Personnel Quality on the Performance of Patriot Air Defense System Operators*, R-3901-A, RAND, 1992.

[j]John D. Winkler and J. Michael Polich, *Effectiveness of Interactive Videodisc in Army Communications Training*, R-3848, RAND, 1990, and Judith C. Fernandez, "Soldier Quality and Job Performance in Team Tasks," *Social Science Quarterly*, Vol. 73, pp. 253–265, 1992.

[k]C. Hammond and S. Horowitz, *Flying Hours and Crew Performance*, P-2379, Institute for Defense Analyses, 1990, and *Relating Flying Hours to Aircrew Performance: Evidence for Attack and Transport Missions*, P-2608, Institute for Defense Analyses, 1992.

As the table shows, the analyses used different measures of output. Sorties are a direct measure of current output—aircraft sorties produced per day. Mission capable rates indicate a capacity, in this case how well a ship is ready to perform its mission, where readiness depends on equipment availability and operability and personnel availability and training readiness. Maintenance downtime is an indirect measure of maintenance output; multitasking indicates the extent to which personnel with multiple-skill training differed from single-skill personnel in their ability to handle different items for repair that arrive randomly. Supervisor ratings are not tied to any specific production process; supervisors were asked to assess a person's productivity at y months (y < 48) in the first term compared with the "typical" productivity at 48 months. Job performance tests measure the capacity to perform a job such as multichannel radio equipment repair, the operation of Patriot missiles, or the operation of a tank.

In every case, the indicators of personnel quality had a positive effect on output. The results for education and AFQT vindicated the services' earlier push to increase recruit quality and confirmed what the services were reporting from field experience: high-quality personnel outperform low-quality personnel. Experience measured by years of service was also positively related to output. Since experience was a factor in all the studies, the robustness of its effect is clear. As shown and discussed below, experience levels rose throughout the 1980s and into the 1990s, increasing military capability.

After the late-1970s crisis, personnel quality improvements were nothing short of dramatic. Table 7.3 shows the percentage of recruits who were high school graduates and the percentage defined as high quality. From 1975 to 1980, the Army recruited less than 60 percent high school graduates at a time when 80 percent of 18–24 year olds had a high school degree. By 1980, only 21 percent of its recruits were high quality. By comparison, consider that the median value of the AFQT score was normed at 50 in 1980 for a nationally representative sample of the youth population. Given that 80 percent of youths were high school graduates, at least 40 percent should have been defined as high quality. The Navy's percentage of high-quality recruits was around the national average, the Marine Corps's was below average, and the Air Force's was well above average.

Table 7.3

Percentage of Recruits That Are High School Graduates and Percentage That Are High Quality, by Service and Year[a]

Service	1975	1980	1985	1990	1995	1997
Army						
High school graduate	58	52	86	94	94	90
High quality[b]	38	21[c]	49	61	64	58
Navy						
High school graduate	74	74	88	90	92	95
High quality	49	44[c]	49	53	60	61
Marine Corps						
High school graduate	47	70	90	93	95	96
High quality	32	35[c]	49	61	62	62
Air Force						
High school graduate	87	84	99	99	99	99
High quality	63	56[c]	67	84	82	77

SOURCE: *Population Representation in the Military Services, Fiscal Year 1997*, pp. D-12 and D-14.

[a]These are all non-prior service accessions.

[b]High-quality individuals are defined as those who score in the upper half of the AFQT distribution and have a high school degree.

[c]Values reflect test misnorming that led to inflated 1977–1981 scores.

By 1985, a turnaround had occurred. In the Army, Navy, and Marine Corps, the percentage of recruits with a high school degree was 5 to 10 percentage points above the national average, and the percentage of recruits that were of high quality, at 50 percent, was also above the national average. These values continued to increase into the 1990s; by 1995, the Army had 94 percent high school graduates and 64 percent high-quality recruits. The Navy and Marine Corps had over 92 and 95 percent high school graduates, respectively, and 60 and 62 percent high-quality recruits. The Air Force also made substantial gains. In 1995, 99 percent of its recruits were high school graduates and 82 percent were high quality. But recruit quality fell after 1995 because of the strong economy and the increase in college attendance, again demonstrating the need for policy action to be responsive to changes in the external environment.

Table 7.4 shows experience trends for enlisted personnel and officers. The large increase in enlisted experience allowed the services to reap substantial benefits from the higher quality of recruits: more expected years of service, higher performance during those years, and, perhaps, a greater increase in performance with experience because high-quality personnel probably learn faster. Average years of service increased by over a third for enlisted personnel and by 10 percent for officers from 1980 to 1995. Enlisted experience reached 7.4 years in 1997, versus 5.5 years in 1980, partly because of the increase in the percentage of recruits with high school degrees and the fact that their attrition was lower. In the early 1980s, the first-term attrition rate was approximately twice as high for non–high school graduates as for graduates. As the services, especially the Army, reached over 90 percent high school graduates, attrition fell and years of service increased.[6] Officers averaged 9.9 years of service in 1980 and 10.8 in 1997, indicating that an additional year of service

Table 7.4

Average Years of Service and Percent Increase in Mean Years for Enlisted Personnel and Officers

	1980	1985	1990	1995	1997
Average years of service					
Enlisted	5.5	6.0	6.8	7.4	7.4
Officers	9.9	9.9	10.4	10.7	10.8
Percent increase since 1980					
Enlisted	—	9	23	34	34
Officers	—	0	5	8	10

SOURCE: *Population Representation in the Military Services, Fiscal Year 1997*, pp. D-17 and D-27.

---

[6]In the late 1990s, Army first-term attrition was once again high, upwards of 35 percent. This was in part due to the Army's tightening of its training standards. Other causes were the more immediate accession of recruits from the delayed entry pool—i.e., recruits who would have dropped out while waiting to enter service instead dropped out during training. Increases in smoking and decreases in physical fitness also played a role.

was gained from these personnel, which is advantageous given their high costs of accession and training and long learning curves.

The gain in quality and experience was driven by systemic changes. Higher pay, bonuses, educational benefits, increased recruiting resources, improved recruiter management, and a recession led to better recruiting and retention. Higher recruit quality and higher retention rates both meant lower attrition, hence less need to recruit. This allowed the recruiting establishment to focus more on attracting high-quality recruits, which fed the positive cycle. These factors worked together fairly rapidly; much of the revolution in recruit quality occurred between 1980 and 1985. Most of the gain in high school graduates had been attained by 1985, and probably over half the gain in high-quality recruits had been attained by then.[7]

The negative cycle at the end of the 1970s was comparably rapid.[8] Both cycles reflected the potential hazards and benefits of an all-volunteer force. The late-1970s cycle destabilized the volunteer force concept, whereas the early-1980s cycle demonstrated the potential for recovery. Both episodes underscore how dependent the volunteer force is on good management, especially in compensation and recruiting. Also, even though both cycles lasted only a short time, their effects on the force quality were long lasting. Just as high-quality recruits produced the benefits described above, lower-quality recruits pointed to less military capability. They had lower pass rates in training, higher attrition and hence more unit turbulence, and lower

---

[7]In 1980, the recorded percentage of high-quality recruits was inflated by an error in test norming. Therefore, the gain in high quality from 1980 to 1985 was actually greater than shown in Table 7.3.

[8]Military/civilian pay fell throughout this period; the GI Bill, an enlistment incentive, was allowed to lapse; and the national economy improved as unemployment fell from 8.5 percent in 1975 to 5.8 percent in 1979. Recruit quality and retention rates sank, reaching perilous lows in 1980. Further, the ASVAB misnorming masked an even lower quality of recruits. The decline in recruit quality fed higher attrition, and higher attrition plus lower retention kept recruiting requirements high. Recruiting resources were not immediately increased, however, so recruiters were hard pressed to make their goals. In a scramble, they shifted more of their effort away from high-quality prospects to the easier-to-recruit segments of the market—non-graduates and those scoring in the lower half of the AFQT score distribution. As a result, although the recruiters largely succeeded in meeting their recruit quantity goals, the proportion of high-quality recruits fell.

scores on tests of skills and knowledge. The services, particularly the Army, tried to thin out low performers by tightening reenlistment standards, but a large portion of lower-quality personnel reenlisted nonetheless.

Is the conventional definition of high-quality enlisted personnel— i.e., having a high school degree and scoring in the upper half of the AFQT score distribution—too narrow? Although these two indicators are associated with high performance, research has shown (Table 7.2) that only on-the-job experience can reveal certain important but previously unobserved aspects of quality, such as effort, reliability, leadership, ability to work as part of a team, and communication skills.[9]

Information about unobserved quality can be inferred from a person's speed of promotion in the first term.[10] If, when education and AFQT are held constant, a person reaches E-4 faster than his or her peers, this might indicate the person is high quality, but it might also simply be a good random outcome. However, a person who reaches both E-4 and E-5 early is probably a high-quality individual. By analyzing promotion to E-4 and E-5, the net effect of the unobserved quality can be detected, although the individual aspects themselves cannot be identified. Empirical work suggests that unobserved quality plays a major role in accounting for the variation of quality among individuals, a role larger than that of education and AFQT.

In addition, quality and later outcomes appear to be linked. Analysis based on the extended measure of quality indicates that those who perform well in their first term stay for longer careers and reach higher ranks that have leadership responsibilities. Putting this in perspective, since education and AFQT contribute to superior performance, it is advantageous to keep these higher-quality personnel. But personnel with high AFQT scores are actually slightly more likely to leave service after the first term than are personnel with low AFQT

---

[9]Along the same lines, SAT scores predict which students will do well at elite colleges but have far less to say about subsequent earnings.

[10]Michael P. Ward and Hong W. Tan, *The Retention of High-Quality Personnel in the U.S. Armed Forces*, R-3117-MIL, RAND, 1985; James R. Hosek and Michael G. Mattock, *Learning About Quality: How the Quality of Military Personnel Is Revealed over Time*, MR-1593-OSD, RAND, 2002.

scores—an empirical fact that has been known for some time. It suggests that the civilian labor market offers somewhat better career opportunities for high-aptitude individuals than does the military, but this is speculation. The good news is that the military tends to keep personnel who perform well on duty, after controlling for AFQT score.[11]

It is worth mentioning several other issues related to the all-volunteer force that may have a place in future, dynamic models for personnel management. While these may not return as prominent issues, they have been of significant concern in the past. Specifically, critics of the volunteer force have questioned its viability given demographic trends, have decried its seemingly perverse impact on the military as a national institution, and have suggested its potential for weakening societal support for the military.

The first of these concerns—that the high accession requirements of an all-volunteer force could not be supported by the decreasing size of youth population cohorts—became moot as accession requirements fell in the early 1980s. The percentage decrease was greater than the projected percentage decline in the youth population from its high point in 1979 to its low point in 1995. The services also recruited more women.

The second concern was that the all-volunteer force would convert the military from "an organizational format that is predominately institutional to one that is becoming more and more occupational."[12] This would affect the legitimacy of and social regard for military service by replacing normative values, such as honor, duty, and country, and "esteem based on notions of service" with "prestige based on level of compensation."[13] Military personnel would no longer define their role commitment and military reference groups in

---

[11]Related work has found that with education and AFQT held constant, personnel expecting faster promotion to E-5 are more likely to reenlist—another sign of pro-selectivity (Richard Buddin, Daniel S. Levy, Janet M. Hanley, and Donald M. Waldman, *Promotion Tempo and Enlisted Retention*, R-4135-FMP, RAND, 1992.

[12]Charles C. Moskos, "Institutional and Occupational Trends in the Armed Forces," in Charles C. Moskos and Frank R. Wood (eds.), *The Military: More Than Just a Job?* Pergamon-Brassey's, New York, 1988, p.15.

[13]Moskos, "Institutional and Occupational Trends."

terms of their military peers, but instead would make comparisons to civilian occupations. The structure of compensation would shift toward salary and bonuses and away from noncash and deferred compensation.

Although the military changed under the all-volunteer concept, there is little evidence it became a weaker institution or a less capable force or was held in lower esteem by society. The volunteer concept eliminated the conscription tax and proved successful in meeting manpower requirements when pay and recruiting resources were sufficient. It is unclear whether military personnel became less likely to define themselves relative to their military role and military peers. Even under conscription, the military was a volunteer force beyond the first term of service for enlisted personnel or beyond the minimum service obligation for officers; the conversion to an all-volunteer force changed the front end. Senior personnel, those volunteering to remain in service, compare their pay and careers to those in the civilian sector as they no doubt did before the all-volunteer force.

The third concern—about an all-volunteer force weakening societal support for the military—stemmed from the observation that the fraction of the youth population serving in the military had declined. Additionally, fewer members of Congress had military experience. Table 7.5 shows enlistment figures for 18-year-olds for 1955 through 2000. In 1960, 1965, and 1970, about one in four young men entered the military. By 1980, about one in seven joined; and a decade after the Cold War, one in 12 joined. The percentage of young women entering has grown since the all-volunteer force began, but at around 2 percent, it remains small.[14] Still, it is not obvious that the decline in the percentage of young men entering the military reduced society's regard for the military or weakened Congress's resolve to keep the military strong. Fewer youth served because the volunteer force succeeded in keeping personnel longer.

---

[14]The table focuses on non-prior service enlistment, but its message would be little changed by including officers. Officers would add about 1 percentage point to the percentage of males entering service and 0.2 percentage point to the percentage of females.

Table 7.5

Enlistment Figures for 18-Year-Olds, 1955–2000

	18-Year-Olds (in thousands)		Non-Prior Service Accessions			Percent Accessing	
Year	Men	Women	Men	Women	Total	Men	Women
1955	1,074	1,068	611[a]	12[a]	623	56.9	1.1
1960	1,323	1,289	381[a]	8[a]	389	28.8	0.6
1965	1,929	1,875	406	8	414	21.0	0.4
1970	1,914	1,868	619	13	632	32.3	0.7
1975	2,159	2,097	374	36	410	17.3	1.7
1980	2,156	2,089	303	49	352	14.1	2.3
1985	1,877	1,809	259	—	297	13.8	2.1
1990	1,849	1,755	193	30	223	10.5	1.7
1995	1,796	1,710	138	29	167	7.7	1.7
2000	2,011	1,918	160	33	193	8.0	1.7

SOURCE: *Population Representation in the Military Services, Fiscal Year 1997*, pp. D-1 and D-10.

[a]Author's estimate.

Economic Theories of Compensation

Military compensation plainly is a key policy instrument for attracting and keeping high-quality personnel and for shaping the personnel force. According to a global survey of pay, "The increasing importance of human capital is transforming pay and the lives of the human resources managers who administer it. Companies see pay as their main tool for recruiting, motivating and retaining good people. All three are important."[15] Another observation: "The true value of a business (or, for that matter, of a household or a country) is often not fully reflected in the audited numbers because markets value assets that don't show up on the balance sheet. What is the key asset not shown on the balance sheet? It sounds too simple, but a good part of it is people. In today's knowledge-based economy, nothing equals the contribution of people."[16] In theory, compensation can be structured in different ways for different purposes.

---

[15]*The Economist*, "A Survey of Pay," May 8–14, 1999, unnumbered.

[16]Michael Milken, "From the Chairman," *Milken Institute Magazine*, First Quarter, 1999.

Military compensation includes basic pay, allowances, special and incentive pay, educational benefits, and retirement benefits.[17] Basic pay now depends solely on rank and years of service. The chief allowances are for subsistence and housing. The sum of basic pay, the subsistence allowance, the housing allowance, and the implicit tax advantage from the nontaxability of the allowances is called *regular military compensation*. Special pay supplements are associated with arduous or dangerous duty—e.g., sea pay, submarine pay, aviator continuation pay, and imminent danger pay (formerly called hazardous duty pay). Incentive pay, such as enlistment and reenlistment bonuses, is paid selectively by military occupation and changes in response to the supply of personnel. Educational benefits are available to enlisted personnel through the Montgomery GI Bill and to officers through Reserve Officers Training Corps (ROTC) scholarships and appointments to the service academies. The services supplement the GI Bill benefits with their "college funds" to attract high-quality recruits for certain military occupations. Retirement benefits do not vest until the 20th year of service, are payable immediately upon retirement from the service, and pay approximately 40 to 75 percent of basic pay depending on rank and years of service at time of retirement. There are many other elements of military compensation—uniform allowances, relocation reimbursement, and so forth—but these are more incidental in nature.[18]

The following subsections describe five theoretical perspectives on, or models of, compensation—general human capital, specific human capital, motivation of effort, initial sorting, and tournaments—and how these relate to military compensation.

General Human Capital.[19] General human capital is assumed to be equally valuable in alternative uses—for instance, as valuable in the

---

[17]A comprehensive description of military compensation and its legislative background may be found in *Military Compensation Background Papers*, 5th ed., Department of Defense, Office of Secretary of Defense, 1996.

[18]In addition, the military provides health care to service members and health care coverage to their families, and attends to the schooling of dependents (e.g., by arranging to provide it directly or by compensating local school districts to allow the children of military families based nearby to attend).

[19]The seminal contributions are Gary S. Becker, "Investment in Human Capital: A Theoretical Analysis," *Journal of Political Economy*, Vol. 70, October 1962, pp. 9–49;

military as in the private sector. Common instances of general human capital are elementary and secondary schooling, and verbal and quantitative skills. Money and equipment constitute investments in human capital, as does the individual's time. The main cost to the individual comes from forgone earnings during the time spent in school or training. Because general human capital can be used anywhere, the employer has no incentive to pay for it, and the worker bears the full cost and receives all the returns. An employer that paid for the investment could not expect to capture the returns without paying the worker a wage above his marginal value product. Such a wage would cause the worker to remain with the employer, but it would also cause the employer to lose money. Therefore, the worker bears all the cost, and the worker's subsequent wage just equals his marginal value product. Thus, in this model, the costs and returns to human capital investment are entirely the worker's, the worker receives a wage equal to marginal value product in each period, and the worker has no incentive to stay with a particular employer.

If education is taken to be an indicator of general human capital, it follows that military compensation must keep pace with market wage for those with education in order to maintain a flow of recruits. One way to judge whether compensation is keeping pace is to compare military pay with the market wage for a particular group. For instance, we can compare regular military compensation for an E-4 with the wage deciles of 22- to 26-year-old white males who are full-time year-round workers.[20] Military pay was at a value equal to the 50th percentile (median wage) in 1982, rose over the next few years to above the 60th percentile, subsided, and then rose sharply to the 70th percentile as the nation's economy slowed down and entered a recession in the early 1990s. The increase in military pay over this period thus fueled the increase in recruit quality. But from 1993 to 2000, military pay fell, relatively, and the services reported increasing difficulty in recruiting.[21]

---

and Jacob Mincer, "On-the-Job Training: Costs, Returns, and Some Implications," *Journal of Political Economy*, Vol. 70, October 1962, pp. S50–S90.

[20]Regular military compensation equals the sum of basic pay, basic allowance for subsistence, basic allowance for housing, and a tax adjustment account for the allowances not being subject to federal income tax.

[21]For further discussion, see Beth Asch, James Hosek, and John Warner, *On Restructuring Enlisted Pay*, AB-468-OSD, RAND, December 2000.

Specific Human Capital.[22] Suppose an employer can profit from a firm-specific investment in a worker if the worker remains at the firm. The only way the employer can be sure to retain the worker is to pay more than the worker can get elsewhere. The employer can do this by using part of the returns to support the increased wage. But on the margin, if the employer shares the returns but pays the full cost of the investment, the investment is unprofitable. A feasible alternative is for the employer to allow the worker to participate in the investment, sharing the returns and costs proportionately. Then the worker's wage is less than his or her opportunity wage during the investment phase and more than the opportunity wage afterwards. The employer pays part of the investment cost and afterwards receives a stream of returns equal to the difference between the worker's marginal value product and wage. Because of the sharing, the worker's wage profile is not as steep as it would be for an equal-size investment in general human capital, where the worker would bear the full cost and receive the full returns.

The specific human capital model suggests why workers at a firm who have greater seniority and who are more educated are less likely to be laid off or dismissed during a business slump. These workers tend to have more firm-specific capital invested in them. Since a worker's marginal value product exceeds his or her wage, there is room for a decrease in the marginal value product before it drops below the wage and triggers a separation. One question this model does not address is: After the investment has been made, what stops the employer from paying the worker less than promised but more than his opportunity wage? An answer stems from the concepts of repeated contracting and reputation that have been developed in game theory. With repeated contracting, the employer who wants the option of making later investments in the worker should not renege on the initial deal. If the employer damages his reputation by acting in bad faith, employees will be unwilling to deal without safeguards on their returns, the costs of which will have to be borne either by the employer directly or by the worker and then passed on to the employer as a cost of the transaction.

---

[22]The seminal contributions here are also from Becker, "Investment in Human Capital," and Mincer, "On-the-Job Training."

In military compensation, bonuses and special pay are mechanisms for rewarding investments in military-specific human capital. They are found in occupations having no direct civilian counterpart—e.g., combat arms, submariner, fighter pilot. Because bonuses and special pay encourage retention, they enable the services to get more return on their investment in military-specific human capital. Bonuses and special pay are also found in military occupations whose skills are transferable to private-sector jobs that offer more than basic pay. Because private-sector wages vary across occupations but the military uses a common basic pay table across all of them, the military must pay some occupations more to make them competitive with outside opportunities.

Motivation of Effort.[23] In this model, the employer wants the worker to exert effort, but effort is assumed to create disutility for the worker. The employer knows that a worker can be induced to exert more effort in exchange for higher future pay, but the employer cannot pay more than the worker is worth to the firm over his/her working lifetime at the firm, given the expected level of effort. Because the worker's choice of effort depends on his/her expected wage growth, the employer offers the worker a schedule of wage/effort profiles. Each worker selects the wage/effort profile that maximizes utility; higher-effort profiles generally have faster wage growth. In addition, to attract the worker, the employer must make the value of the selected wage profile at least as great as that of any alternative opportunity with a different employer. But faster wage growth cannot go on forever, because it would exceed the worker's value to the firm. Hence, if the firm is to offer steeper wage profiles to induce greater effort and thereby make the worker more valuable to the firm, the firm must also stipulate a mandatory retirement date for each wage profile.

In the military, pay growth occurs through time in service, promotion, and the year-to-year increase in expected future retirement benefits. At a given rank, the growth of time-in-service basic pay is shallow and offers little incentive for greater effort. In fact, during the first 10 or so years of service, most wage growth comes from promo-

---

[23]A key initial source is Edward P. Lazear, "Why Is There Mandatory Retirement?" *Journal of Political Economy*, Vol. 87, December 1979, pp. 1261–1284. Lazear's later reflections appear in "Personal Economics: Past Lessons and Future Directions," *Journal of Labor Economics*, Vol. 17, No. 2, pp. 199–236.

tions. But promotions are not the main focus of the effort motivation theory, which aims at effort on a given job—i.e., in the military, at a given rank. For personnel who reach 10 years of service, expected retirement benefits become an increasingly important factor in their compensation. Also, discounting causes the expected present value of retirement benefits to rise dramatically as the 20th year of service approaches. For many service members, the value of annual compensation, including the increment in expected future retirement benefits, is greater than the value of his or her output during these years—a result that fits the ideas in the incentive motivation model. The military also sets a mandatory retirement date of 30 years of service—another fit with the model.

But do expected retirement benefits actually offer a strong incentive for performance during service years 10 to 20? To reach 20 years of service at current rank, the service member must guard against a mistake or misbehavior that would result in demotion or dismissal from service. Passable behavior probably does not require much effort, so, from an organizational perspective, it is important that people *not* reach 10 to 12 years of service without having been selected for their knowledge, skills, initiative, and effort. This observation suggests the value of sorting and tournament models.

Initial Sorting.[24] Despite the information gained from résumés, job interviews, screening tests, and even handwriting analysis, employers are hard pressed to tell if a new employee is well matched to the job. Job changes are frequent among young workers; about half of all unemployment falls on workers under age 25. To some people, frequent job changes represent aimless churning within the labor market; to others, they are a sorting process. Economic models of sorting assume there is uncertainty about a worker's ability, even given observable characteristics such as education, test scores, and prior experience. Each worker has an incentive to claim high ability, especially if the employer cannot verify the claim. What the employer must do, then, is observe the worker's performance on the job and be willing to offer a contract in which higher future wages are conditional on revealed ability. Using repeated observations comparing

---

[24]Boyan Jovanovich, "Firm-Specific Capital and Turnover," *Journal of Political Economy*, Vol. 87, December 1979, pp. 1246–1260.

the worker's performance to external or internal standards, the employer can assess the worker's ability. Since assessment comes in the context of performance within the firm, ability becomes synonymous with "quality of job match."

The military does a great deal of sorting. Most officers have an initial service obligation of four years, and much of the sorting occurs after this period and during the rank of captain (O-3). That is, most officers (70 to 90 percent) continue in service after completing their service obligation but then leave service at a fairly steady rate while they are captains, say, years six to 11. About 50 percent of officer cohorts reach the O-4 promotion window around 11 to 13 years of service, and about 40 percent continue to 20 years of service (retirement eligibility). Among enlisted personnel, about 30 of every 100 accessions are lost to first-term attrition, and about half of the remaining 70 reenlist. So, about 35 percent of those in an enlisted cohort stay for a second term, and eventually about 13 percent reach 20 years of service. Those who stay tend to be personnel who have performed well relative to their peers and personnel who have a taste for the military.

Tournaments.[25] Tournament models analyze competition for promotion within a hierarchical organization. The organization is assumed to be a pyramid: the number of positions decreases as job rank increases. The employer wants the most-capable workers to ascend to positions of highest authority because high-level decisions affect productivity at all lower levels. The tournament model describes a mechanism to induce effort and sort workers efficiently. Consider a firm with two ranks, mid and high, and a worker at the mid level. The worker's incentive to compete for promotion to the high level depends on the probability of promotion and the high-level wage. The product of these two is the worker's expected wage at the high level. If the worker is a wealth maximizer, the expected wage must exceed the mid wage or the worker has no incentive to compete.

Now add a low level to the organization. A low-level worker's incentive to compete for promotion to mid level depends on the increase

---

[25]The concept of tournaments comes from Sherwin Rosen, "Prizes and Incentives in Elimination Tournaments," *American Economic Review*, Vol. 76, September 1986, pp. 701–715.

in expected wealth, which depends on the mid-level wage and the wage expected with promotion from mid to high level. Also, the low wage must be lower than the mid-level wage to create a positive incentive to compete. Now assume that the probability of promotion from mid to high level is less than that of promotion from low to mid level. This assumption accords with the job pyramid, but it is subtle because the probability of promotion depends not only on the number of higher-rank positions, but also on the rate of outflow of personnel from those positions and the firm, which in turn depends on the promotion and wage structure. It follows that the ratio of the high-level to the mid-level wage must be greater than the ratio of the mid-level to the low-level wage.[26] In other words, the wage structure is skewed by rank. The tournament model thus implies that wages should be disproportionately greater at each rank. Moreover, competing for promotion may entail exerting greater effort than otherwise and investing in human capital. In most cases, the workers compete against one another, not an external standard. To move up the promotion queue, the worker must outperform fellow workers, who are also exerting more effort. If the competition is stiff, some workers will not bother to compete. Also, if highly able workers can get by with little effort, they will do so unless they face high stakes at the higher ranks in the form of skewed pay.[27]

The military makes extensive use of promotions. Competition for them tends to induce greater effort among the competitors, and since competition favors the more able, they are the ones more likely

---

[26]The mid wage cannot be greater than the high wage. If it were, there would be no incentive for mid-level workers to compete for the high-level job. For low-level workers, the increase in expected wealth depends on mid- and high-level promotion probability and wages. Conceivably, the mid wage could be *less* than the low wage if the expected high-level wage were high enough. The second assumption rules out this possibility. This assumption is consistent with a rationality constraint: If the mid-level job required higher skills or more effort than the low-level job, qualified workers being hired from outside the firm would not choose the mid-level job unless it paid more than the low-level job.

[27]Another model is that of principal and agent (e.g., see Bengt Holmstrom, "Moral Hazard and Observability," *The Bell Journal of Economics*, Vol. 10, No. 1, Spring 1979, pp. 74–91). Typically, an efficient principal-agent contract involves the principal assigning a property right to the agent (e.g., a portion of net revenues or stock options), with the principal remaining as a residual claimant. This model does not translate easily to the military context because it is difficult to assign a property right to income or the opportunity for added wealth (stock option).

to compete. Because each rank in the military is a stepping stone to the next, the expected gain from future promotions enters current decisionmaking about whether to compete and how much effort to exert. At first glance, military compensation does not appear skewed, because basic pay does not have a prominent nonlinear increase with rank. But as mentioned earlier, after 10 years of service, the draw of retirement benefits grows much stronger each year. Since retirement benefits depend on rank at retirement, the gain associated with promotion is substantial. Thus, retirement benefits impart skew, the more so with promotion.

This does not mean retirement benefits are necessarily the best way to skew the wage structure. One could argue that a better way would be to move a large portion of military compensation from retirement benefits to current compensation to create a skewed basic pay structure.[28] And there is a definite limitation associated with using retirement benefits to create skew. The services operate their promotion systems in a way meant to provide equal opportunity for advancement regardless of one's military occupation. As a result, the expected pay grade at retirement is roughly the same across occupations. This affords little freedom to tailor career paths to specific occupations, as might be advantageous in the case of certain skills and areas (e.g., information technologists and acquisition specialists). By the same token, pay differentials across occupations can be introduced only through bonuses or special pay, neither of which counts toward increasing the size of a service member's retirement benefits.

ISSUES FOR THE FUTURE

The capability of future military forces depends on the versatility and leadership of their personnel, which, in turn, depend on the quality of not only the enlisted and officer recruits, but also the policies affecting training, experience, career development, and cohesion.

How can the military recruit and retain these high-quality personnel? The history of the all-volunteer force tells a profound story. Competitive, well-structured compensation and adequate recruiting re-

---

[28]See Beth J. Asch and John T. Warner, *A Theory of Military Compensation and Personnel Policy*, MR-439-OSD, RAND, 1994.

sources have enabled the military to attract and keep the personnel it most needs, whereas lapses in compensation and recruiting led to a manpower crisis. Moreover, a willingness to conduct formal experiments and demonstrations, as well as an ongoing program of research and analysis, was required to determine an appropriate level and structure of compensation. By managing the volunteer force well, policymakers increased quality, increased reenlistment, and increased the average career length—all of which contributed to greater military capability. The types of tools used successfully in the past can be applied to future policy alternatives. Moreover, the scope of inquiry can expand, since most past analysis focused primarily on supply issues rather than on demand (requirement) issues.

New recruiting challenges have already presented themselves. More high school graduates are choosing to attend college, shrinking the traditional recruiting pool. The number of new accessions needed has risen from levels held low during the drawdown and may rise further if force size is increased to ease the manning of peace operations, as some suggest. Private-sector wages have been rising faster for persons with some college than for those with only a high school degree, and faster for those with high aptitudes. Also, officer retention has fallen in certain areas (e.g., Air Force captains). Close monitoring and timely action are required to keep military careers competitive with private-sector opportunities, and innovative recruiting strategies are needed to align the military's recruiting strategies with the increased numbers of young people going to college and to then penetrate the two-year college market. The risk of not taking these steps is a costly loss of high-quality human capital and leadership.

Officers and enlisted personnel can expect the heightened pace of peacetime operations to continue. Service members complain of being too busy and that the unpredictability of deployments disrupts family life. Even so, those who have been deployed have somewhat higher reenlistment rates than those who have not, although extensive deployment into hostile areas can eat into this higher rate.[29] Thus, the military needs to manage the tempo of work life and the nature and frequency of deployments for all personnel.

---

[29]James R. Hosek and Mark Totten, *Does Perstempo Hurt Reenlistment? The Effect of Long or Hostile Perstempo on Reenlistment*, MR-990-OSD, RAND, 1998.

The services will also want to look at longer careers in occupations with high military-specific capital. Today's pay table is based on rank and time in service, providing most pay growth through promotions. Senior military specialists may not need a higher rank (and the command authority it provides), however, although they presumably do need incentives for performance. The pay table may have to be restructured to provide these incentives. Specialists in other areas, such as information technology, may require more-creative work environments with access to the latest technology.

Future manpower requirements will also be an issue. As future force concepts, such as the cyber soldier and information warrior, are developed, manpower will have to be brought into defense planning so that force structure assessments consider the benefits and costs of having more-experienced personnel in certain functional areas or occupations. Changes in experience mix, pressures to maintain or increase recruit quality, and private-sector wage growth may lead the military to reevaluate the structure of military compensation and military careers with respect to their cost-effectiveness in attracting, keeping, motivating, and sorting personnel.

Chapter Eight

ADAPTING BEST COMMERCIAL PRACTICES TO DEFENSE[1]

Frank Camm

Over the last decade, the Department of Defense (DoD) has sought increasingly to transform its basic approach to warfighting and the methods it uses to support warfighters.[2] As part of this effort, leaders and influential observers of DoD have repeatedly encouraged DoD to emulate "best commercial practices" (BCPs)—the practices of commercial firms that have been recognized by their peers as being the best among firms engaged in similar activities. Over the past 20 years, many successful firms have found that BCPs offer an important new source of information for improving their competitive position. In particular, they have used information on BCPs to complement and even replace more traditional forms of analysis associated with organizational innovation. Properly used, BCPs provide a rich

---

[1]Much of the material in this chapter draws on empirically based policy analysis that I have conducted at RAND with John Ausink, Laura Baldwin, Charles Cannon, Mary Chenoweth, Irv Cohen, Cynthia Cook, Jeff Drezner, Chris Hanks, Cynthia Huger, Ed Keating, Brent Keltner, Beth Lachman, Jeff Luck, Ellen Pint, Ray Pyles, Ken Reynolds, Susan Resetar, Hy Shulman, Jack Skeels, and, particularly, Nancy Moore, co-leader with me of the Project AIR FORCE project on new approaches to sourcing and contracting. Together, we have reviewed the empirical literature on identifying and implementing best commercial practices in logistics, environmental management, and sourcing-related processes. We have conducted detailed case studies on practices in over 60 commercial firms and many DoD organizations. I could not have written this chapter without the work we have done together, but I retain full responsibility for the chapter's content.

[2]For more information on this effort, see U.S. Department of Defense, *Quadrennial Defense Review Report,* Washington, DC, 30 September 2001; U.S. Congress, General Accounting Office, *Defense Reform Initiative: Organization, Status, and Challenges,* GAO/NSIAD-99-87, Washington, DC, April 1999.

database on new ideas that have worked in particular settings and on the factors underlying their success. Advocates of using BCPs in DoD argue that BCPs could serve a similar role in helping the department transform itself. That is, BCP assessment could complement other methods used to support transformation, including traditional forms of public policy analysis.

By contrast, skeptics argue that the institutional setting of DoD (and, more broadly, the federal government) is so different from the settings of commercial firms that BCPs have little to teach DoD. As my colleague, Gregory Treverton, argues, "The public and private sectors are alike in all the unimportant respects." Differences in basic values, incentives, constraints, and operating environments, as well as DoD's profoundly political setting, limit the applicability of BCPs observed in commercial firms.

This chapter describes how DoD can use BCPs to help transform activities that have appropriate commercial analogs, particularly activities in the defense infrastructure.[3] These include administrative services, generic business and personnel services, education and training, sourcing, and the elements of base operations, medical care, information services, logistics, and civil engineering that are separable from direct military operations.[4] Even where appropriate commercial analogs exist, fundamental differences between the public and private sectors require that BCPs be carefully tailored for adaptation to DoD needs. DoD has already found ways to tailor such BCPs to its peculiar needs and can do a great deal more of this in the future.

This chapter first addresses the general challenge of adapting BCPs to DoD. It then uses sourcing BCPs relevant to DoD to illustrate more concretely how to adapt BCPs to DoD's peculiar institutional setting.

---

[3]For simplicity, I speak of a monolithic DoD in search of BCPs. Of course, DoD must rely on its constituent components and agencies to find and adapt BCPs in ways that work best for them. When I speak of DoD taking any particular action, I mean a particular decisionmaker or activity within DoD acting within the constraints relevant to that part of DoD.

[4]DoD can use commercial models for other activities as well, but finding useful analogs becomes increasingly difficult as the department's core military activities are approached.

The initial overview covers four topics: (1) what BCPs are and how DoD can benefit by learning more about them; (2) the close relationship between many BCPs and operational total quality management (TQM)—i.e., TQM absorbed into day-to-day operational decision-making; (3) how DoD can identify BCPs relevant to its needs; and (4) how DoD can use formal change management techniques to adapt such BCPs to its own goals and operational environment.

Using BCPs associated with the general commercial practice called "strategic sourcing," the chapter then walks through a sequence that uses change management to adapt BCPs to DoD's needs. This sequence identifies strategic goals relevant to DoD sourcing policy, sourcing BCPs relevant to these goals and the way in which these BCPs relate to one another, DoD's recent efforts to adapt these BCPs and key barriers to more complete adaptation, and tactics DoD can use to mitigate these barriers. DoD could use a similar sequence to pursue adaptation of BCPs in any policy area.

WHAT IS A BEST COMMERCIAL PRACTICE?

BCPs are typically tied to processes—i.e., activities that transform inputs into outputs in any organization. Processes can, for example, transform strategic priorities into requirements, development resources into new products, or labor and material inputs into serviceable parts. BCPs occur in processes that use fewer inputs to yield better or more outputs faster. They make specific processes or their outputs "better, faster, and/or cheaper."

Most BCPs are more likely to occur in firms that do business with one another rather than with DoD. So, to find BCPs, DoD must typically look well beyond its traditional horizons.

Examples

Caterpillar was among the first firms recognized for world-class logistics performance. In 1990, it could fill 98 percent of all requests for parts within 48 hours, anywhere in the world. During the Persian Gulf War, Saudi Caterpillar tractors, supported by Caterpillar, were available throughout the war, whereas U.S. Army Caterpillar tractors, reliant on organic DoD support systems, were not. The specific changes in logistics and sourcing processes that brought about such world-class performance in order fulfillment and reliability are BCPs.

In 1986, the Emergency Planning and Community Right-to-Know Act (EPCRA)[5] created the toxic releases inventory (TRI), which identifies the physical volume of a list of toxic chemicals emitted by major U.S. companies each year. The TRI made many corporate executives aware, for the first time, of how their firms were affecting the environment (and how much money they were wasting as emitted chemicals). Under a variety of voluntary programs, firms committed themselves to cut emissions by an order of magnitude over three years. By implementing operational and environmental BCPs tailored to their particular industrial processes, they met their commitment without having to cut industrial production. DoD cut its TRI emissions by 50 percent over three years. It could very likely go even further.

From 1993 to 1997, AMR, the parent company of American Airlines, cut the cost of all its purchased goods and services by 20 percent (relative to inflation) without affecting performance levels. Honda cut similar costs by 17 percent from 1994 to 1997 without a performance loss. Both firms already had sophisticated purchasing programs in place. They introduced BCPs, tailored to their priorities, into their purchasing and supplier management processes to achieve these improvements; further improvement has continued in both firms as they have refined their approaches. DoD has sought savings of this magnitude primarily in incremental A-76 cost-comparison studies.[6] AMR and Honda's experiences suggest the virtues of using a much broader approach focused on process change.

Commercial Practice: Neither Monolithic nor Easy to Define

A BCP is not a *specific* best way of doing something that can be picked up and moved anywhere else. On the contrary, the first lesson

---

[5]42 U.S.C. 11001-11050, also known as Title III of SARA (Superfund Amendments and Reauthorization Act).

[6]Office of Management and Budget (OMB) Circular A-76 defines a process DoD can use to compare the costs of using public and private sources to execute a particular work scope.

from commercial practice is that the private sector accommodates an extraordinary variety of policies and practices.

What works in one place may not work in another. The best commercial firms are always trying to learn from other firms with comparable processes, but they recognize that they cannot simply emulate another firm's practice without understanding why it works. Firms look for what might work in their own setting, and when they adopt a BCP, they may use it in such a new way that it is hard to tell exactly what was learned. The fact that best practices must be *adapted* to each new setting makes them elusive; even if all commercial firms applied best practices, there would be considerable variation among them.

Some firms stand out as particularly innovative. The best firms' practices differ precisely because these firms have found different—better—policies and practices that other firms have difficulty emulating.

To complicate things further, BCPs do not stand still. The variation observed in the commercial sector reflects informal experiments that constantly test the effectiveness of doing things differently. As particular practices succeed in appropriate settings, these practices become, by definition, BCPs. They prevail as long as they continue to yield success when appropriately applied. Over time, constant innovation, imitation, and competition yield variations that work even better, constantly displacing practices once considered the best. Firms that use BCPs do not identify specific practices that they then rely on indefinitely so much as they continually seek the best practices available. A commitment to always continue this search underlies the BCPs in the best firms.

Why DoD Should Care

Global competition is driving the best commercial firms to improve every aspect of their businesses. These firms are learning how to (1) determine requirements more quickly and with greater precision to increase the firm's agility and reduce its waste, (2) integrate organic processes across functional boundaries and align those processes with the needs of customers,[7] (3) build relationships with outside sources that integrate those sources with organic processes and the buyer's customers, (4) manage these relationships to realize the buyer's and seller's expectations, and (5) pick sources that can do this.

DoD faces such challenges itself, but in a different environment. Commercial firms face competition that threatens their survival day by day. DoD operates in a more lethal environment, but one that is truly threatening only during a contingency. When DoD actually projects military force, it can expect to face opponents with access to much of the information about BCPs that DoD has. In a global setting, global commercial practices can be adapted anywhere.

Because commercial firms are learning how to perform in an increasingly turbulent, unpredictable, competitive environment, the BCPs they develop often give particular attention to managing uncertainty more effectively. With care, DoD can use BCPs to ask how better to manage the increasingly uncertain environment in which it operates. Care is required to distinguish risks whose root causes are similar for both DoD and commercial firms (such as technological innovation and the behavior of external sources) from risks unique to DoD (such as those driven by immediate military concerns or congressional politics).

As these examples illustrate, DoD can expect not only cost reductions, but also improvements in process performance and product quality and reliability.[8] What aggregate improvements BCPs will allow DoD to achieve cannot be predicted; they will depend on the particulars of the processes and products DoD addresses and on how it adapts specific commercial practices to its own institutional setting. But it is reasonable to expect that wherever DoD can find com-

---

[7]*Functions* are communities of specialists trained in similar skills, such as maintenance, financial management, or contracting. They provide skills to many different processes. *Processes* use specialists from different functions to deliver a product, such as a repaired part or a meal, to a customer. Functions tend to have an input-oriented perspective; processes tend to have an output-oriented perspective.

[8]DoD's goals reach well beyond concerns about cost, performance, quality, and reliability. As we shall see, DoD's institutional environment places high value on other goals for sourcing policies and practices. Given the priorities captured by that environment, however, BCPs are most likely to help DoD improve its costs, performance, quality, reliability, and so on. We return to this point below.

mercial analogs, it should be able to achieve significant and continuing improvements that outweigh the effort required to achieve such change. DoD can learn over time what to expect and thus specifically where to continue investing.

A BCP That DoD Uses Today: Lean Production

The book *The Machine That Changed the World* [9] did much to popularize the relevance of BCPs. It detailed how the Japanese automobile industry was outperforming its North American counterpart, which, in turn, was outperforming the European counterpart. The term *lean production* was used to explain why the performance of these industries differed, what North American industry could learn from the Japanese automobile industry, and what some BCPs were in 1990.

What did Japanese firms do right? They cut design, marketing, and production cycle times to catch errors quickly, thereby generating less wasted effort than North American firms did. They built quality into the cars to reduce the cost of after-the-fact inspections and rework, thereby slashing parts and work-in-progress inventories, which, in turn, reduced the capital investment required to produce cars. They used basic TQM tools to understand what customers wanted and then aligned all of their design, marketing, and production processes to give customers exactly that.

When RAND analysts first examined lean production, they expected to find nothing useful to DoD. What could an organization focused on warfighters learn from firms driven by commercial accountants? The answer depends on how one looks. Lean production pays particular attention to uncertainties associated with customer demand, production process performance, and the performance of external sources. When RAND analysts looked at the automobile industry, process by process, and compared the uncertainties in these processes with those in analogous processes in military logistics, they discovered that lean production offered exactly what DoD needed to

---

[9] By James P. Womack, Daniel Roos, and Daniel T. Jones, MacMillan/Rawson Associates, New York, 1990.

deal with its logistical uncertainties.[10] RAND helped the Air Force develop its first application of lean production: agile combat support (originally called "lean logistics"). The Army followed shortly thereafter, developing its velocity management, and the Marines then developed their precision logistics. Each adaptation is tailored to its specific setting in DoD.[11] None would have occurred the way it did had DoD not discovered how to learn from the experiences of the Japanese automobile industry.

What About Best Government Practice?

People in DoD tend to seek best practices elsewhere in DoD or at least in other agencies of the federal government. Benchmarking efforts in the federal government tend to focus on the government, and they suggest that cross-government learning is important.

Nevertheless, DoD ought to look as far afield as commercial firms outside DoD's traditional orbit, because these firms are experiencing a real revolution in business affairs that is likely to continue. Moreover, while defense spending is growing, it still makes up just a few percent of the U.S. economy. All else being equal, any innovation is thus many times more likely to occur outside the defense sector than within it. The likelihood is even higher for processes that are more commercial than military in character, especially those processes that make up DoD's infrastructure. If DoD can monitor commercial ideas that succeed, it can cull the best ones for its own use without having to experience the failure that some ideas, inevitably, will meet with.

In effect, BCPs help DoD focus its leadership and in-house innovation efforts on its core missions, because BCPs give it access to useful

---

[10]I. K. Cohen, John B. Abell, T. Lippiatt, *Coupling Logistics to Operations to Meet Uncertainty and the Threat (CLOUT): An Overview*, R-3979-AF, RAND, 1991; Timothy L. Ramey, *Lean Logistics: High Velocity Logistics Infrastructure and the C-5 Galaxy*, MR-581-AF, RAND, 1999.

[11]For example, agile combat support increasingly emphasizes the importance of adapting to new contingencies quickly so that the Air Force can become more "expeditionary"—i.e., can project force more quickly and reliably in a contingency. Velocity management emphasizes improvement in peacetime performance to bring Army logistics processes under control to make them more reliable and less costly.

external information about its noncore activities—i.e., those with close analogs in the commercial sector. If most technological and organizational innovations outside its core activities are highly likely to occur elsewhere, why should DoD waste its own effort and resources keeping these activities at the cutting edge of capability? BCPs offer an alternative approach, one that DoD can use if it creates and maintains core, in-house capabilities for adapting BCPs. These new capabilities would free resources and leadership efforts to focus on core activities.

That said, when an innovation enters DoD (or the rest of the government), from whatever source, it is fair game for adaptation. Adaptation of any best practice is first and foremost about change management. Much of the following discussion on implementing change applies as well to best *government* practices as to BCPs.

OPERATIONAL TOTAL QUALITY MANAGEMENT AND BCPs

As often as not, BCPs involve the application of TQM in particular settings. The ideas at the core of TQM are simple and logical: TQM offers a straightforward approach to using reliable empirical evidence to track and adjust management decisions. *Operational* TQM does this as an integral part of day-to-day management. It does not maintain a functional distinction between a "quality community" and line management.

Key Benefits of TQM: Links Between Customers and Processes, and Continuous Improvement of Resulting System

TQM seeks clear, empirically based answers to three fundamental sets of questions:[12]

- Who are the firm's customers and what do they want? If the firm has many customers, do they want different things? What can the firm do to reflect these differences in its products?

---

[12]For an eloquent demonstration of this point, see Arnold S. Levine and Jeff Luck, *The New Management Paradigm: A Review of Principles and Practices*, MR-458-AF, RAND, 1994.

- What processes does the firm use to serve its customers, and how does it coordinate these processes so that they all align with customer needs? If the firm buys inputs from external sources, how does it align these sources with its customers' needs?

- How can the firm improve? Is it serving the right customers? How can it improve its understanding of its customers' needs? How can it better coordinate and align all processes it uses to serve its customers?

Described this way, TQM sounds like simple common sense. To a large extent, it is. But such common sense focuses management's attention on (1) its customers rather than its internal constituencies (which can legitimize themselves under TQM only by serving the ultimate customer), and (2) cross-functional processes rather than the internal communities in functional divisions (which are thereby deemphasized).

TQM rejects internal standards as a basis for monitoring internal performance in favor of challenging all process owners to look for ways to improve the standard performance of each process. The logical place to look is at similar processes used elsewhere—BCPs, for example.

TQM relies on the idea that these things do not happen unless individuals do them. TQM aligns the behavior of individuals with an organization's goals. It formally recognizes the importance of stakeholders and the fact that they must participate in any process management. It then seeks rules of engagement, roles and responsibilities, metrics, and incentives that help them work toward a common purpose. In particular, it helps stakeholders involved in any process look beyond the process and understand how it can contribute to the goals of the organization as a whole. Such "common sense" demands that an organization that wants to change not only recognize what that change means for the job behavior of the individuals affected, but also prepare careful plans directly focused on how the change will affect the behavior of the individuals. These ideas demand that behavioral change lie at the heart of any organizational change induced by a decision to adapt a new practice.

TQM is most commonly associated with continuous improvement—*kaizen*, in Japanese. *Kaizen* can mean taking a given process or customer base and improving it over time—perhaps by shortening the process's cycle time or improving the firm's understanding of the customer base's latent desires. In these cases, change tends to come from the bottom, because only the individuals who understand a process's subtleties can offer useful process improvements.

More dramatic "reengineering" and strategic changes demand that an organization smash its existing processes and even seek new markets. Both of these require top-down initiative, because only by seeing the organization as a whole can one imagine a qualitatively different direction for its internal processes or its customer base. This is not what most people associate with TQM, but such distinctions may be overdrawn. In this chapter, TQM refers to the management of changes of all kinds that improve an organization's understanding of its customers, improve the processes used to serve those customers, or prepare individuals in the organization to do these things in a way that promotes the organization's goals. This definition is probably broader than the one normally used.

Implementing *kaizen* as a standard element of day-to-day management is a major change that requires commitment and sustained support from the top. Management systems throughout the organization must change; there is no *kaizen* way to implement *kaizen*. Conversely, the specific changes that implement reengineering and alternative strategic direction must come from somewhere (even if from a consultant, someone inside the organization must decide to pay attention). Radical changes to a part of an organization appear incremental to the organization as a whole—*kaizen* on a higher level. TQM provides useful principles for creating the culture that generates new ideas, be they large or small.

The best examples of the close relationship between TQM and BCPs come from looking at the ISO 9000 management standard and the U.S. Department of Commerce Malcolm Baldrige Award.[13] For both, third-party auditors use a checklist to verify that an organization

---

[13]ISO 9000 is a family of voluntary standards created and maintained by the International Standards Organizations (ISO). Organizations can be certified to ISO 9000 if they maintain a detailed set of management processes. The U.S. Congress in 1987 created the Baldrige Award to recognize U.S. organizations that best exemplify a detailed list of characteristics based on the principles of TQM.

complies with specific requirements. The lists differ, but both include the key elements of TQM and seek to know how far TQM has been integrated into day-to-day management. Both also seek to understand an organization's customers and processes, how it ensures and improves quality, and how it enables and motivates people.

DoD could use such checklists. They work best when applied by a third party, which can provide objective discipline as well as the services of a knowledgeable consultant. In fact, third-party auditors could help DoD verify whether it has adapted the critical elements of any specific BCP it might consider.

TQM Viewed with Great Suspicion by Many in DoD

Some people have heard so many different versions of TQM that it just confuses them. As quality consultants split hairs over the fine points in trying to differentiate their wares so as to gain market share, the central message gets lost in the noise. In fact, as explained above, the core ideas in TQM are simple and logical.

Other people have participated in formal efforts to implement TQM in their agencies. They have heard all the rhetoric but have seen no more performance improvement from these efforts than from their predecessors. What they remember is the decline in budget and manpower authorizations made ahead of any savings that TQM would provide, and that they lost capability when the savings did not come. Commercial firms have similarly learned that TQM works only when it goes from being a stand-alone program to an operational part of day-to-day management, which happens less often than it should. TQM must be operationally implemented to overcome this problem.

Still others, particularly those looking for new commercial ideas, observe that the term *TQM* is now largely gone from the business press. If fact, where successful, TQM has been absorbed into day-to-day management. As implementers routinely use internal management audits and align their internal organizations to the processes they use to serve customers, they mention TQM less and less. Yet ISO 9000 (and its clones[14]) has become the dominant standard that good firms use to qualify suppliers. Profitability aside, the Malcolm Baldrige Award remains the dominant U.S. measure of high management performance. Operational TQM is now so pervasive in commercial firms that it is no longer seen as separate from good management.

IDENTIFYING BCPs

To identify BCPs, an organization must learn how to look beyond its own boundaries in a new way. It must learn how to exploit the increasingly abundant information resources available on other organizations' practices. The World Wide Web is a logical first place to look for references to appropriate information, if not for the information itself. Three types of resources can be helpful:[15]

1. *Professional associations and the conferences, courses, and publications they sponsor.* These offer ideal places to track rapidly evolving BCPs and their effects. They provide information on concepts and cases, as well as a natural place to meet specific practitioners and consultants who can provide additional information about practices in specific settings.

2. *Books and journals.* The major accounting firms maintain databases on best practices and have begun to publicize their data in books.[16] More generally, books and journals offer detailed examples of best practices from an industry or academic perspective. The industry perspective typically includes hands-on detail critical to understanding the usefulness and transferability of a specific practice. The academic perspective typically puts cases in a broader context, one that is particularly useful in balancing the often over-optimistic message of the industry press. (Companies

---

[14]The automobile, aerospace, and telecommunications industries have now developed their own variations on ISO 9000: QS-9000, AS-9000, and TL-9000.

[15]Table 8.1, shown later, provides examples of items in these categories that RAND used in its work on strategic sourcing.

[16]For example, Arthur Andersen offered a comprehensive approach to adapting BCPs in Robert Hiebeler, Thomas B. Kelly, and Charles Ketterman, *Best Practices: Building Your Business with Customer-Focused Solutions,* Simon and Schuster, New York, 1998.

rarely voluntarily share information on their failures to adapt best practices.) Both perspectives are useful.

3. *Benchmarking.* A specific way to place an organization in a broader context, benchmarking typically proceeds in three stages. Stage 1, high-level benchmarking, identifies the general nature of BCPs. The resources described above focus at this level. Stage 2, quantitative benchmarking, identifies a set of metrics and compares them across organizations with similar processes. This can occur in specific studies or on an on-going basis in benchmarking networks of organizations with similar processes. Stage 3, practitioner benchmarking, allows an organization to send the people who will adapt a new practice to a place where the practice already works so they can talk to the people who make it work. Face-to-face comparisons of day-to-day operations on the best practitioner's site give an organization's own practitioners latent knowledge about how they will have to change to successfully transfer the practice that they can get in no other way.

Described in this way, the identification of BCPs is a natural aspect of market research. DoD has always conducted market research but has typically thought about it only in terms of acquisition and contracting. By contrast, all parts of the best commercial firms are becoming increasingly aware of what is happening outside their firms. Because BCPs can benefit DoD so widely, market research should interest all process managers, notably those responsible for requirements determination, organic process design, design of relationships with external sources, source selection, and ongoing performance management for internal and external sources.

Adapting BCPs for Use in DoD

The most challenging aspect of adapting a BCP is successfully transferring it into DoD—i.e., ensuring that DoD changes in a way that allows it to benefit from the BCP. BCP adaptation is thus first and foremost about managing organizational change. Any new practice will require a formal program, the goal of which will be to adapt the practice to the DoD setting and to adapt standing DoD policies and systems to support the new practice. These change-related activities present a serious challenge, no matter where the change comes from.

Formal change management systematically addresses a series of issues that are also relevant to adapting best practices:

- Who in DoD does the change affect? Do they have the same relative importance in DoD as their counterparts in the commercial world?

- Who has to change their behavior on the job? What are the best ways to induce such changes in DoD relative to the commercial source organization?

- What DoD management systems must be synchronized with the change? How do they compare to analogous management systems in the source organization?

Given the close relationship between adaptation of BCPs and change management, as well as the fact that any adaptation must ultimately support and induce a specific organizational change, DoD can best approach the challenge of adaptation itself through the lens of change management. If DoD plans for successful change management, it will also, by definition, define an effective adaptation. The two fit hand in glove.

Successful management of organizational change is in itself a BCP. As DoD adapts more commercial practices, change can become easier. In the meantime, however, change is likely to be harder in DoD than in the best commercial firms. Personnel can keep this in mind as they review material about BCPs and think about how such practices will change when they come to DoD.

Every large, complex organization faces a diverse constellation of stakeholders. Large commercial firms typically identify their shareholders, customers, employees, suppliers, and the outside community as the stakeholders relevant to their success. DoD serves taxpayers, warfighters, and military families rather than shareholders and customers. Its employees are organized differently and have different rights. It is subject to much greater external scrutiny and pressure than a typical commercial firm is; political constituencies, working through Congress or more directly, are particularly important. Because key stakeholders shape any large organizational change, DoD can expect them to alter BCPs as these practices move from any commercial organization to DoD. Such pressure could well reduce the net benefits that a BCP offers, as DoD services each stakeholder's needs.

Still, DoD can affect the balance among these stakeholders to some extent. In particular, many BCPs shift control from a functional community to a customer. For example, in a logistics setting, BCPs emphasize the availability of parts to a final user (the customer) over the efficiency of the transportation system (the functional community) used to deliver these parts. DoD has demonstrated that it can do this in selected settings.

Once DoD acts to shift authority from its functional communities to its ultimate users, DoD can accept the presence of its key stakeholders and maintain an open process in which they negotiate to shape any new commercial practice. But to succeed, DoD must guard against allowing any one stakeholder to capture the adaptation process.

Change is only as effective as the senior leadership support it can garner and sustain through the full change effort. So a successful change is typically as large as the senior leadership supporting it allows. Adaptation of a BCP may look like a large change in DoD simply because of differences between the DoD and best commercial settings. This makes senior support and proper sizing all the more important.

A common approach seeks support for a small change in one locality and uses empirical evidence of benefits from that change to engage a higher, broader level of leadership. Even with high-level support from the beginning, an incremental approach is likely to sustain the support required to make a large change and to limit the risks associated with such a change. Ways to do this include the following:

- Use carefully instrumented pilots to test BCPs and adapt them to DoD as appropriate.

- Initially choose BCPs that require the smallest adaptation to transfer them to DoD.

- Make BCPs as compatible as possible with the local DoD culture where they are received. Use waivers where they are available.

- Refrain from introducing BCPs too close to DoD's core combat-related activities so that any failures will have no more than limited effects on DoD and on the change effort itself.

Caution is important, but a proposed change must be big enough to get people's attention.

Structural Differences Between DoD and Most Best Commercial Firms

Corporate America and the major elements of DoD initially learned from one another about how to harness the myriad energies of a giant enterprise to a single purpose. Well into the 1950s, they used similar management methods to plan and operate their organizations. Since the 1950s, however, and at an accelerating rate since the 1970s, the best commercial firms have moved away from the organizational form that DoD continues to favor.[17] In particular:[18]

- While the military services and defense agencies favor strong centralized structures, best commercial firms have reduced their corporate headquarters staffs and devolved authority.

- While DoD organizations rely heavily on strong functional organizations (such as logistics, civil engineering, and financial management), best commercial firms increasingly align themselves along process lines associated with products that cut across such functional lines.

- While DoD favors clearly defined rules, roles, and responsibilities over motivation and incentives, best commercial firms rely relatively more on formal incentives to align employees with the

---

[17]The characteristics attributed to DoD here, of course, apply to the federal government as a whole.

[18]These are not the only factors that create differences between the public and private sectors. For example, DoD tends to draw a greater distinction between effectiveness (such as military capability) and efficiency (cost) goals than commercial firms do, because DoD is subject to many laws and regulations that do not affect private firms. And while DoD emphasizes procedural openness, fairness, and fraud prevention, best commercial firms are more pragmatic in their management of processes and of fraud and abuse.

organization's goals. These firms give their employees more discretion than DoD does.

These differences are real and significant, but they should not be given more importance than they warrant. DoD organizations will probably sustain their preference for directing many actions from the center and using their functional communities to do this. They will also rely more on specific rules than on incentives to align local activities with the center, and they will continue to emphasize openness, accountability, equity, and integrity. DoD must adapt any BCP to make it compatible with these priorities.[19]

To adapt a BCP, DoD must verify that relevant practices that work with commercial management systems also work with analogous DoD systems. Organizationwide management systems differ in all organizations. Because of the growing difference between DoD and the best commercial firms, the differences relevant to four types of management systems are particularly important in this context:

- *General information systems and, in particular, internal transfer prices and related decision-support information.* Prices work in qualitatively different ways in DoD and typical commercial firms.

- *Incentive and motivation systems.* Every organization has its own approach to motivating workers.

- *Workforce management systems.* DoD's need to deploy forces and its heavy reliance on a labor force with little lateral entry create workforce management challenges that the best commercial firms do not face.

- *Systems to release excess resources, particularly labor.* The federal Office of Personnel Management (OPM) and federal unions place constraints on DoD that differ from those faced by the best commercial firms, even those with strong unions.

Planning for major organizational change raises many issues similar to those associated with planning for military action. Before DoD ex-

---

[19]How best to adapt a BCP always depends on the context. The last part of this chapter illustrates this with sourcing examples.

ecutes a military action, it typically develops a detailed contingency plan. As the first day of a military campaign approaches, DoD incorporates more and more details about the actual situation to make the execution as well coordinated and free of surprises as possible. Surprises will occur, however, and the plan must adapt repeatedly as the action proceeds. But the initial plan provides enough structure to anticipate surprises and have resources in place to respond when they occur. Very similar statements could be made about how DoD develops a new weapons system, major end item, or subsystem.

DoD can think about large organizational changes in a similar way. DoD knows how to identify risks and plan against them; it can use this knowledge in planning how to adapt a BCP to its own cultural setting. Key elements of such a plan are likely to include

- A clear, succinct statement of goals, with metrics to characterize the goals. DoD can use these goal-oriented metrics to negotiate adjustments as change proceeds.

- A way to break the change into a simple, defensible set of chunks compatible with the degree of support available for the change.

- For each change, a list of the behaviors that must change on the job, barriers to these changes, and plans to overcome each barrier. These plans address appropriate roles and responsibilities, training and other resource needs, and milestones.

- A coherent endgame that ensures the change is integrated with all appropriate DoD-wide management systems and is sustainable at the end of the transition.

The natural tool to use for this monitoring is a Shewhart cycle,[20] a variation of which is shown in Figure 8.1. Such a cycle provides an integral part of a quality management system that keeps senior leadership well informed about a change's status.

---

[20]Named for Walter A. Shewhart of Bell Telephone Laboratories, it is also popularly known as a PDCA, or Plan-Do-Check-Act, cycle.

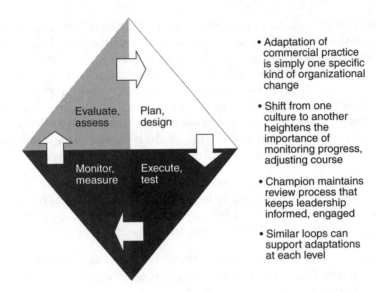

Plan, design

Evaluate, assess

Monitor, measure

Execute, test

- Adaptation of commercial practice is simply one specific kind of organizational change

- Shift from one culture to another heightens the importance of monitoring progress, adjusting course

- Champion maintains review process that keeps leadership informed, engaged

- Similar loops can support adaptations at each level

Figure 8.1—A Continuous Improvement Loop Indicating Whether Adaptation Is on Track

The champion (i.e., person responsible for the change) uses the cycle to plan and design the elements of a new commercial practice, execute and thereby test this practice in DoD, monitor the test and measure the practice's performance against the stated goals of the change, and evaluate the outcomes.[21] If need be, the champion adjusts the design of the commercial practice and executes it again in the test setting. The champion includes the coalition supporting the change—i.e., the stakeholders whose organizations must alter their behavior for the change to succeed—in this loop. This gives the coalition an opportunity to approve or redirect as well.

---

[21]At RAND, we associate such a cycle with "maturation." For a discussion of maturation in the context of technological innovation, see J. R. Gebman, H. L. Shulman, and C. L. Batten, *A Strategy for Reforming Avionics Acquisition and Support,* R-2908/2-AF, RAND, 1988.

An Illustrative Example: Strategic Sourcing as a Basket of BCPs

Strategic sourcing links an organization's sources for the goods and services it uses as inputs to its corporate strategic goals. Typically, an organization identifies its customers' needs and then verifies that its sources of goods and services are tightly aligned with those needs. Every commercial firm that uses strategic sourcing customizes it to its own setting, but broadly speaking, a firm that uses strategic sourcing

- Identifies its customers' needs and translates them into measurable metrics.

- Organizes its internal processes to choose and manage sources, internal or external, accordingly.

- Develops relationships with key high-quality sources that become partnerships enabling both buyer and seller to benefit from pursuing the needs of the buyer's customers.

- Uses metrics that reflect customer priorities to measure the performance of the high-quality sources that the firm partners with.

Both buyer and seller benefit by working jointly to improve performance; joint continuous improvement lies at the heart of these partnerships.

The strategic goals relevant to strategic sourcing in DoD can be summarized as follows:[22]

- Improve military capability

- Sustain or improve safety and quality of life

- Reduce total ownership cost[23]

---

[22]These closely mirror the set—efficiency, equity, and integrity—proposed in Steven Kelman, *Procurement and Public Management,* AEI Press, Washington, 1990. Kelman inferred these by examining the rationale underlying federal procurement regulations.

[23]"Total ownership cost" is a commercial measure of all the costs associated with an activity or asset over its lifetime that DoD has begun to use. It covers all the costs of "owning" an activity or asset.

- Honor socioeconomic commitments

- Sustain the openness, equity, and integrity of the sourcing process.

With regard to military capability, the commercial analog is the value of output as measured in monetary terms. Because no one metric exists to measure military capability, this connection between the value of output and a source of inputs is more difficult to make in DoD than in the commercial sector. As for safety and quality of life, both DoD and commercial firms are concerned, but DoD is far more so. The safety of flight in a high-performance combat aircraft, for example, presents a far greater challenge than does safety assurance in most activities occurring in commercial firms. And quality of life more often applies to the workplace in commercial firms than to entire communities, as it does in DoD.

Both DoD and commercial firms also seek to cut costs. But commercial firms have far better cost accounts than DoD does and can more easily pursue comprehensive estimates of cost, such as the total ownership cost. DoD cost accounts are not even good enough to meet the standards that the Defense Contract Administration Agency (DCAA) requires of private-sector suppliers to DoD.[24] DoD needs better cost accounting procedures to go along with BCPs; perhaps they can be imported together.

DoD faces more-challenging socioeconomic goals than any commercial firm does, but commercial firms still have such goals, some self-imposed and some imposed by government regulators.[25] DoD can learn from how the best commercial firms service their socioeconomic goals, but by and large DoD will find its own way.

Finally, DoD faces more-challenging procedural openness, equity, and integrity goals than most commercial firms do. Again, DoD can

---

[24]Standard government cost accounts are typically too incomplete to allow a third party to audit them. They are also not well structured for linking total government costs to outputs or for reflecting how changes in work scope affect costs. Government accounts focus on tracking the application of congressional appropriations, not on the levels of cost relevant to management decisions.

[25]Of particular relevance to sourcing is the fact that many firms maintain goals to use small and disadvantaged businesses as sources.

learn from the way commercial firms pursue these goals. But DoD must be sensitive to its differences from the best commercial firms and verify that the best practices it imports can be adapted to the DoD setting. For example: Formal public-private competitions are far more common in DoD than in the commercial sector, because they provide the openness and equity that the federal setting demands.[26] But formal competitions can accommodate many of the sourcing practices that the best commercial firms routinely use.[27]

Under these circumstances, it may be easiest for DoD to focus its search for attractive BCPs on process changes that can enhance its capability, safety, and cost goals. It can then reflect its socioeconomic and administrative process goals as constraints inherent in the DoD institutional setting; any BCP that can enhance capability, safety, or cost goals must be compatible, when adapted, with DoD's operational socioeconomic and administrative process constraints to be useful to DoD.[28]

BCPs Relevant to DoD's Strategic Goals

Table 8.1 lists examples of the sources RAND analysts drew on to identify BCPs relevant to DoD.[29] The professional organizations

---

[26]Public-private competitions allow public and private sources to compete, in special forms of source selection, for selected government workloads. Private firms rarely use formal competitions to choose between organic and contract sources.

[27]For example, DoD can use noncost factors to compare sources, limit comparisons to preferred providers, and reward successful sources with extended contracts. The discussion below provides more detail.

[28]Such a distinction between goals and constraints does not imply that one is more important than the other. It is a natural—in fact, a necessary—part of any effort to evaluate a BCP when multiple goals or performance attributes are important.

[29]This table and the discussion here draw heavily on Ellen M. Pint and Laura H. Baldwin, *Strategic Sourcing: Theory and Evidence from Economics and Business Management,* MR-865-AF, RAND, 1997; Nancy Y. Moore et al., "Commercial Sourcing: Patterns and Practices in Facility Management," PM-667-AF, RAND, 1997; Frank Camm and Nancy Y. Moore, "Acquisition of Services in 2010: Ideas for Thinking About the Future," internal document, RAND, 1999; Nancy Y. Moore, Laura H. Baldwin, Frank A. Camm, and Cynthia R. Cook, *Implementing Best Purchasing and Supply Management Practices: Lessons from Innovative Commercial Firms,* DB-334-AF, RAND, 2002; John Ausink, Frank A. Camm, and Charles Cannon, *Performance-Based Contracting in the Air Force: A Report on Experiences in the Field,* DB-342-AF, RAND, 2001; Laura H. Baldwin, Frank A. Camm, and Nancy Y. Moore, *Federal Contract*

Table 8.1

Sources Used for Ideas on BCPs

Professional organizations, conferences, and courses	National Association of Purchasing Managers International Facility Management Association Building Owners and Managers Association Council of Logistics Management
Books	Timothy M. Laseter, *Balanced Sourcing: Cooperation ad Competition in Supplier Relationships,* Jossey-Bass for Booz-Allen & *Hamilton,* San Fran cisco, 1998
	Jordan D. Lewis, *The Connected Corporation: How Leading Companies Win Through Customer-Supplier Alliances,* Free Press, New York, 1995
	John Gattorna (ed.), *Strategic Supply Chain Alignment: Best Practice in Supply Chain Management,* Gower Publishing, Aldershot, England, 1998
	Ricardo R. Fernandez, *Total Quality in Purchasing and Supplier Management*, in Total Quality Management Series, Saint Lucie Press, FL, 1994.
Journals	*Harvard Business Review* *Sloan Management Review* *Int'l Journal of Purchasing and Material Management* *Supply Change Management Review*
Benchmarking	Arizona State University Center for Advanced Purchasing Studies Michigan State University Individual exemplar firms

shown represent commercial professionals interested in purchasing and supplier, facility, building, and logistics management. All such groups give sourcing BCPs a great deal of attention in their meetings, research programs, conferences, and courses. Arizona State and Michigan State, shown as benchmarking sources, both maintain well-known programs of research on best practices.

These sources suggest many specific BCPs for DoD to consider. A number of them are listed and defined here, after which their recent

Bundling: A Framework for Making and Justifying Decisions for Purchased Services, MR-1224-AF, RAND, 2001.

status in one part of DOD (the Air Force) and the potential for expanding their application in DoD are described.[30] Table 8.2 summarizes this information. Keep in mind that DoD cannot use all of these BCPs and that it must tailor those it does use for its own needs.

The specific BCPs that DoD might consider include the following:

- *Core competencies.* Identify those capabilities critical to an organization's future success or its raison d'être. These core competencies constitute an organization's unique value-added and hence cannot be outsourced. (That said, note that very few commercial firms outsource everything that lies outside their core competencies.)

- *Chief purchasing officer.* Elevate the CPO to the position of executive-level champion for purchased goods and services. CPOs own the processes that the organization uses to reach make-or-buy decisions, choose specific external sources, and manage relationships with these providers. Commercial CPOs generally do not make such decisions themselves.

- *Metrics.* Use metrics for make-or-buy decisions, source selections, or source management that promotes organizationwide, strategic goals.[31]

- *Total ownership cost.* Measure effects on cost using TOC to monetize as many factors as possible and apply them to organizationwide goals. TOC tends to allocate overhead costs to specific sourcing decisions to reflect all the direct and indirect costs relevant to a decision. Specific TOC measures are best tailored to the capabilities of an organization's cost accounts.

- *Multifunctional teams.* Develop sourcing policy decisions using multifunctional teams composed of members that have been (1) relieved of other duties, (2) trained in team processes, and

---

[30]The status of these practices in the Air Force is current as of late 2000.

[31]Such a change can have much broader effects than might first be apparent. For the Air Force, for example, it completely reframes the Air Force's current approach to determining requirements for many infrastructure activities.

Table 8.2

Summary of Openings for and Barriers to Air Force Adaptation of BCPs

BCP	Status in Air Force Today	Barriers to Further Adaptation
Identify core competencies	Does this now	Processes for choosing competencies are not aligned with Air Force strategic goals
Appoint executive-level CPO	Has this	Not effectively empowered to build strategic sourcing policy across functional lines
Use organizationwide metrics in sourcing	Is comfortable with metrics	Functional metrics not properly aligned with Air Force–wide goals
Apply total ownership cost in sourcing	Is moving in this direction	Current accounts do not support it; definition unclear
Use multifunctional teams in sourcing	Uses IPTs	Teams not yet empowered or incentivized to transcend functional priorities
Stratify supplier base	Is moving in this direction	No clear barriers but no clear metrics to reveal value easily
Use simplified acquisition	Is moving in this direction	Contracting does not consider full effects
Buy services in larger bundles	Is moving in this direction	Small business rules strongly discourage this
Use substitutes for competition	Faces strong opposition to this	CICA and small business rules strongly discourage this
Use nonprice criteria to choose sources	Is moving in this direction	Sourcing processes still require a price criterion
Reduce number of suppliers	Faces strong opposition to this	CICA and small business rules strongly discourage this
Consolidate contracts to improve leverage	Is experimenting with corporate contracts	Current data systems do not support "spend analyses" required to do this
Use performance-based statements of work	Is moving in this direction	Still learning what it means and how best to do it
Use higher skilled personnel	Is moving in this direction	Training is hard; so is handling personnel who cannot be trained

NOTE: IPT = integrated process team; CICA = Competition in Contracting Act (1984).

(3) empowered to act for their functions without consultation. The reward structures for team members should reflect the performance of the teams the members work on with respect to organizationwide goals.[32]

- *Stratified supplier base.* Use strategic criteria to stratify the supplier base. Suppliers of high-value inputs that are critical to the buyer's performance or that present other significant risks should be managed with greater care and by higher-skilled staffs than should suppliers of low-value inputs of a more generic character that present fewer risks.

- *Simplified low-priority buys.* Use automation and purchase cards to simplify low-priority buys. Automation releases personnel focused on transaction management; purchase cards further reduce transactions costs, particularly when bundled with auditing and reporting support from issuing banks.

- *Larger bundles.* Buying bundled services can allow the buyer to benefit from provider economies of scale and scope. They can also reduce transaction costs, particularly when the buyer devolves responsibility for oversight of many services to the provider.

- *Substitution of benchmarking and TQM standards for formal competition.* Benchmarking and TQM standards promote continuous improvement and make the external world more visible to the buyer. They can yield comparative information about capabilities, on a continuing basis, that buyers traditionally could only get from formal or "yardstick" competitions. By contrast, repeated competitions can impose unnecessary administrative costs and discourage long-term, joint innovation.

- *Less reliance on price.* In source selections, rely less on price and more on nonprice selection criteria. Nonprice factors can be critical to understanding total ownership costs and a source's ability to reduce them over time.

---

[32]Such a change has broad implications. It lifts decisionmaking out of a functional frame and tends to accelerate any process that depends on input from multiple functional communities (e.g., requirements determination). Inputs traditionally provided in series now occur simultaneously, with feedback from all players rather than just those downstream in a decision process.

- *Reduced number of providers.* The best buyers have cut their number of providers by an order of magnitude. Reduce the number used and select the survivors using such standards as ISO 9000 or data on past performance. Deeper investments can then be made in the remaining sources to promote joint innovations and match specific providers more effectively to emerging needs.

- *Consolidated contracts.* Consolidating contracts with remaining providers can reduce transaction costs and simplify deeper, strategic investment in a provider. It can also improve the buyer's leverage with the seller by highlighting the value of its total buy from the seller.

- *Performance-based statements of work and objectives.* Write performance-based rather than process-based statements of work and objectives—i.e., tell a provider what to do, not how to do it. This forces the buyer to think more carefully about what it values and gives providers more latitude to innovate.

- *Upgraded skill levels in purchasing organizations.* As strategic purchasing and supplier management policies grow in importance, they can no longer be managed in a back office separate from the firm's core interests. Upgrading can be paid for by simplifying small acquisitions.

In pursuing useful BCPs, DoD should not view this list simply as a menu of items it can mix and match arbitrarily. The best commercial firms find that these practices work best as an integrated package. The presence of one raises the effectiveness of the others, for several reasons:

1. Strategic sourcing relies heavily on high-level interest and carefully structured incentive systems. The latter cannot succeed without appropriate metrics. Effective buyer-seller partnerships require everyone's cooperation, and that takes support from the top.

2. Workforce upgrades are easier when funds are available from sourcing efficiencies. Automation and simplification can free up sourcing personnel. A buyer can use the savings to upgrade re-

maining personnel so that they may then plan and manage more-complex and more-creative sourcing relationships.

3. Performance-based statements of work succeed only when buyers can trust sellers enough to reduce process-oriented oversight and let providers exercise enough discretion to exploit performance-based statements of work. The right source must be in place before performance-based criteria can be used.

That said, DoD need not adopt all the suggestions to realize benefits from any one of them. Instead, DoD could recognize these synergies and verify that the mix it picks generates enough of them. This is a special challenge if DoD breaks the introduction of strategic sources into pieces to be introduced sequentially. Such a strategy would affect the realization of important synergies.

Key Barriers to DoD's Adaptation of Sourcing BCPs

DoD is already introducing some aspects of the BCPs identified above, but it has not been as aggressive about any of these practices as the best commercial firms have. In some cases, goal differences account for the differences in practice; in others, DoD can emulate BCPs much more closely. Recent Air Force experience illustrates these points:

Core competencies. DoD and the Air Force are well aware of the concept of a core competency and have begun to use it in their planning. Sourcing reviews associated with defining "core" depot activities, Defense Reform Initiative Directive (DRID) 20, and recent OMB policy based on the Federal Activities Inventory Reform (FAIR) Act of 1998 have forced DoD components to think more carefully about their core missions.[33] Unfortunately, in doing this they have relied heavily on the organic functions that currently provide services. The best commercial firms do not go this route; they handle such policy

---

[33]Congressional policy on depot use requires DoD to define the "core workload" relevant to its organic depots. DRID 20 required DoD to identify all manpower positions that could be considered for potential outsourcing via public-private competition. OMB's use of the FAIR Act requires DoD to put out for formal competition a prescribed fraction, which grows over time, of the manpower positions it has available for potential outsourcing.

at a higher level to avoid conflicts of interest with current internal providers.

Chief purchasing officer. The Air Force has a CPO, but the position lacks the authority held by CPOs in the best commercial firms. The Air Force CPO lives primarily within the acquisition community; commercial CPOs are more closely aligned with the line activities that use purchased goods and services, which gives them greater authority to work across functional boundaries in pursuit of broad, strategic organizational goals.

Metrics. Metrics of all kinds pervade the Air Force, but they tend to be designed and collected within functional organizations to meet their immediate needs. For instance, financial management focuses metrics more on managing against a plan than on responding to the needs of warfighters or their families. By contrast, BCPs explicitly align their metrics with customer needs.

Total ownership cost. DoD has been directed to start measuring TOC, using life-cycle cost as a basis. The quality of DoD cost accounts limits this effort by making it hard to trace all costs to the sourcing decisions they should influence.

Multifunctional teams. Integrated process teams (IPTs) that include members from all functions supporting a process are now a routine part of the Air Force and the rest of DoD. But these multifunctional teams are not used the same way best commercial firms use theirs. DoD team members rarely get the training on team processes that commercial team members receive, they cannot commit their functions to a decision without consultation, they are rarely managed and evaluated against specific organizationwide goals, and their members are not rewarded on the basis of such evaluations. Functional structures and the career patterns associated with them remain much more structured in DoD than in the best commercial firms, so DoD's functional organizations exercise relatively much more authority.

Stratified supplier base and simplified low-priority buys. The Air Force is moving toward stratified acquisition, which uses standard, generic contract terms to handle routine purchases and builds customized relationships with sources for strategically important inputs. Simplified acquisition and purchase cards are cutting the workload of contracting personnel associated with small transactions. A more commercial approach could reduce burdens on the functional personnel who use purchasing cards. Lightning Bolt 99-2 is a policy reform initiative that, among other things, selectively uses highly skilled teams to address complex new acquisitions of support services.[34] Overall, this effort would probably yield larger gains if the Air Force managed it against Air Force–wide goals, such as TOC, rather than metrics tied to each specific initiative.

Larger bundles. The Air Force is moving toward bundling activities and outsourcing them together. It has initiated several large, multi-functional cost comparisons for base-level services. Recent Small Business Administration (SBA) regulation requires that any federal agency bundling previously unbundled services must document the benefits that will accrue; it also limits the benefits that can be used to justify bundling.

Substitution of benchmarking and TQM standards for formal competition. The Competition in Contracting Act (CICA) of 1984 makes it hard to limit the use of competition for external-source selection. Additionally, congressional legislation and OMB Circular A-76 require that the use of public-private competition continue. DoD will have to rely heavily on competition until these directives change. Under acquisition reform, however, the Air Force is using award terms and other techniques to extend the period between competitions.

Less reliance on price. The Air Force relies increasingly on best-value competitions to choose external sources for services. These competitions all place heavy emphasis on past performance and often consider other nonprice factors. But regulations require that price remain a significant selection criterion.

Reduced number of providers. CICA limits any effort to reduce the number of sources considered in a competition or to allow offers by invitation only. But acquisition reform now allows the Air Force to "down-select" during a source selection in more or less formal ways. A down-select effectively reduces the range of competitors to those

---

[34]U.S. Air Force, SAF/AQ, *Aerospace Acquisition 2000,* April 23, 1999, available at http://www.safaq.hq.af.mil/acq_ref/bolts_99/bolt2.htm (as of October 22, 2002).

most likely to meet the government's needs. Thus, it can focus on a smaller field of offerors as it shapes the final version of any work statement.

Consolidated contracts. Contract consolidation is expanding in DoD. The Defense Logistics Agency has been writing so-called corporate contracts for over a decade, and the Air Force has several pilot corporate contracts in place and is seeking additional candidates. DoD continues to experience great difficulty in its attempts to consolidate contracts across DoD contracting organizations and across organizational lines within a provider firm.

Performance-based statements of work and objectives. The Air Force has initiated what are, in effect, over 20 successful pilots of performance-based statements of work during the last two years. This experience has revealed that knowledgeable, motivated acquisition personnel can write such statements of work in a DoD setting without much difficulty. But training remains a problem, and many noncontracting functionals and customers believe that such an approach presents more risks than rewards.

Upgraded skill levels. The Air Force strategy for contracting anticipates a smaller, more highly skilled contracting labor force. The Air Force is moving this direction but is still unclear what to do with personnel who cannot be upgraded. The Air Force has not yet extended this strategy to noncontracting personnel important to service acquisitions.

Insights from Commercial Experience on Overcoming Key Barriers

Looking across these BCPs, a number of barriers appear again and again, highlighting the importance of finding ways to ameliorate them. These include barriers to appointing an effective CPO, developing relevant metrics, using multifunctional teams or simplified acquisition, and defining requirements and performance-based statements of work. Less obviously, agreements negotiated a long time ago with competition advocates or small business advocates give them effective veto power. Some of these agreements are now reflected in laws and regulations. BCPs that can avoid these difficulties are easier to implement than those that cannot.

Most of these problems are not unique to DoD or the federal government. The best commercial firms have faced and found ways to deal with most of them. Their experience suggests that how an organization approaches strategic sourcing is often as important as what elements of strategic sourcing the organization pursues. Table 8.3 sketches the possibilities, which, in effect, illustrate how principles discussed earlier apply to the implementation of strategic sourcing BCPs.

DoD can do three things to address function-related barriers to strategic sourcing:

1. DoD can measure change in terms that transcend functional boundaries and reflect DoD-wide goals. The best commercial firms use "billets eliminated," not just "billets reviewed for potential outsourcing," to measure progress; and they use comprehensive measures of cost, not the number of items procured through a new form of contract. Such metrics are performance oriented: they tell leaders and workers what matters to the organization, not necessarily how to make detailed changes. Such an approach would encourage DoD organizations to measure costs better, thus

Table 8.3

Possible DoD Approaches to Strategic Sourcing

BCPs for Effecting Organizational Change	Implications in DoD
Use metrics relevant to parties affected to support change, justify investments, measure ultimate success, support incentives	Cost savings, billets eliminated; develop baseline, accounts that can measure these "accurately enough"
Build a coalition of parties involved	Unit commander, functionals, contracting, other support functions
Frame change to degree of senior support available	Within a major command or function and at a single base; keep as simple as possible
Have organization designated a special pilot	Attracts resources, allows policy waivers
Incentivize the parties involved	Awards, resources retained, performance reviews, protection for displaced personnel
Train personnel affected as a team	Substance of change, support tools, team process

making it easier to justify investments to support change, and would reward organizations that best promote DoD sourcing goals.

2. DoD can verify that an appropriate group of leaders, at the right level, supports change and can therefore form the core of the coalition used to plan and manage change. The coalition would include not only manpower (for A-76 actions) and contracting, but also relevant functional providers and customers of the services in question. Rapid turnover in leadership and current DoD team methods complicate coalition formation. But change metrics based on DoD-wide goals can help any group of leaders or team quickly understand the usefulness of change and make appropriate adjustments as the change evolves toward completion. In particular, such metrics can assist ultimate customers in understanding how sourcing actions can help them.

3. DoD can focus initially on smaller changes that require changes in only one organization. For example, it can pursue new sourcing practices at one base or in one functional area, but not both. As experience accumulates, an initial change can be used to build the case for broader change *if* the initial change anticipates settings for future changes and collects data relevant to future settings. Small changes limit the number of leaders who must coordinate their efforts to effect change; they also increase the likelihood that change can be completed during the leaders' limited tours of duty together.

Pilot programs are well suited to this approach. DoD has provided waivers that release many of the constraints discussed above in selected locations. Although such waivers are hard to get across the board, they can be used to establish selected beachheads, which, in turn, can supply the evidence that DoD can use to revisit the constraints.

Although performance metrics and incentives must be linked for change to be effective, there is wide scope to link performance metrics to whatever incentive system is compatible with an organization's corporate structure. DoD could use metrics like those used by the best commercial firms without changing its own incentive system much—as long as performance measures affect the incentives that DoD normally applies to the people who must change their behavior.

For example, cost-cutting goals heavily drive DoD sourcing policy. DoD can use measures of total operating cost that are as similar to those used in BCPs as its own cost accounts will allow. But it must find its own way to reward those who succeed in cutting cost. It might allow a successful organization to retain a portion of the cost savings, even if DoD needs the dollars saved more elsewhere. Or it could prominently reflect cost savings achieved in the performance reviews of the personnel involved, and use this information to affect future promotion, training, and other career management decisions.

DoD uses training to explain how people must change their behavior to make implementation successful. The best commercial firms typically use a broader approach to training, including material on effective team processes, problem solving, and the change process itself. Such training is most successful when it engages the people who will have to work together as a team to effect change (for sourcing, for example, people in contracting, manpower, and the relevant functionals, as well as the people who consume the services in question) and uses case materials tailored to the particular change in question. The case materials should reflect both specific socioeconomic and procedural factors relevant to DoD sourcing and details of commercial practice that help explain its success in the private sector.

Taken together, the BCPs discussed above point to the potential for large-scale, continuing change in DoD. It is important to remember that a similarly rich set of BCPs could be identified for practically every aspect of DoD's infrastructure activities. If DoD pursues all of these, it will enter a state that the best commercial firms increasingly take for granted: one of continuing change in which personnel have to learn to accept ongoing adjustment as a normal part of their day-to-day activities.

Change is already moving so fast and on so many fronts in DoD that many of the personnel whose behavior must change no longer understand how the changes are supposed to fit together or how to set priorities when they do not fit. These personnel do not even know who to go to for answers. Unless this state of affairs ends quickly, continuing efforts to change will overwhelm DoD personnel with "innovation fatigue" and leave them disillusioned about the possibility of progress. Unfortunately, DoD's constantly changing environment does not allow it the luxury of slowing its own change efforts. It has no choice but to learn how to live with continuing change. As DoD learns to knit together coherent packages of DoD-relevant metrics, leadership, pilots, incentives, training, and so on for each particular set of BCPs it considers adapting, it will also need to learn how to knit these packages into larger and larger programs of change. The commercial ideas offered here about how to implement individual sets of BCPs can also help DoD think about effecting change on a broader scale.

PART III. NEW TOOLS FOR DEFENSE DECISIONMAKING

INTRODUCTION TO PART III

RAND and other analysts have developed and refined a number of techniques for coping with uncertainty and making decisions that will have consequences over years, even generations. These techniques might be thought of in three broad categories: exercises, strategic products, and "groupware."

Drawing on earlier work in Europe, war-gaming for U.S. military planners was developed at the Naval War College in the late 19th century as a way of "getting into the minds" of potential military adversaries in order to develop and test alternative operational strategies.[1] In war games, the flow of events is affected by and, in turn, affects decisions made by players representing more than one "actor" or "side" that relies on less-than-perfect information. Simulations are different. Here, players represent only one actor, and some events may be determined before the game is played. Players can have incomplete and possibly misleading information—based on what sensors and human intelligence happen to provide—or they can be assumed to have complete and accurate information.

Both war games and simulations are referred to, in shorthand, as exercises. Such exercises can be conducted with educational or analytic objectives in mind: typical educational objectives include learning new lessons, reinforcing old lessons, and evaluating the understand-

---

[1]For further detail, see Peter P. Perla, *The Art of Wargaming*, Naval Institute Press, Annapolis, MD, 1990. Also see Peter P. Perla and Darryl L. Branting, "What Wargaming Is and Is Not," *Naval War College Review*, September/October 1985; and Peter P. Perla, "Games, Analyses, and Exercise," *Naval War College Review*, Spring 1987.

ing that has been gained; typical analytic objectives include developing strategy, identifying new issues, building consensus, or setting
priorities. Exercises exist to test and refine human interaction, not to
calculate outcomes. They explore decisionmaking by forcing players
to make decisions; they achieve value by producing qualitative assessments of decisions that are made and not made and the effects of
those decisions.

Exercises are generally

- Based on scenarios—i.e., credible, internally consistent, scripted
  events that set the scene and scope for players (although exercises can also be used to develop scenarios).

- Tolerant of some oversimplification and artificial assumptions.

- Designed by people acknowledged as experts in a particular area.

- Guided by rules and procedures to assure the logical flow of
  cause and effect.

Exercise play is most fruitful when participants free themselves from
the constraints of "conventional wisdom" and suspend disbelief,
much as they would when reading a well-written work of fiction.
Getting the most out of games requires exposure of participants to a
structured process of post-game analytic feedback. A well-structured
exercise whose players have relevant experience and good information should yield insights about

- The feasibility of strategies, as well as their strengths and weaknesses.

- Key factors or variables that drive the results.

- The sensitivity of the results to variations in the factors or variables.

Strategic products, the second broad catgory of techniques, help
evaluate the broader implications of changes in the planning environment. They come in two forms, *strategic planning*—"the evaluation and choice processes that determine how the world will be
viewed, and the goals that the organization will pursue given this

world view"[2]—and *strategic forecasting*. Underlying both of these is scenario-based planning.

Scenario-based planning was pioneered in the late 1960s by Royal Dutch Shell, virtually the only major oil company to anticipate the changes that would occur in the oil market in 1973–1974. Now a standard technique for long-range planning in the face of uncertainty, scenario-based planning rests on the premise that the future cannot be predicted accurately enough for good planning. Scenario-based planning rarely aims to predict; it is "a tool for ordering one's perceptions about alternative future environments in which one's decisions might be played out."[3] As one leading practitioner puts it, "The point is not so much to have one scenario that 'gets it right' as to have a set of scenarios that illuminate the major forces driving the system, their interrelationships, and the critical uncertainties."[4] Trends and key uncertainties are used to establish a range of "futures" well enough to define a manageable number of plausible, internally consistent scenarios.[5] Planners then insert themselves into each future environment and assess how their near-term decisions affected the long-term futures.

In assessing these effects, planners use common and adaptive strategies. *Common strategies* are largely independent of which specific future is anticipated; *adaptive strategies* are developed early and then executed later only if specific variations of the future ensue. Because adaptive strategies tend to be costly and risky to execute, they are developed only for selected future circumstances. And because some futures are more desirable than others, some strategies seek to improve the future while others concentrate on how to cope with the less desirable ones. In this way, defense planners help identify ac-

---

[2]Paul R. Kleindorfer, Howard C. Kunreuther, and Paul J.H. Schoemaker, *Decision Sciences: An Integrative Perspective*, Cambridge University Press, New York, 1993, p. 236.

[3]Peter Schwartz, *The Art of the Long View*, Doubleday, New York, 1991, p. 4. Also see Peter Schwartz, *The Art of Long View—Paths to Strategic Insight for Yourself and Your Company*, Bantam Doubleday Dell Publishing Group, Inc., New York, 1966.

[4]Paul J.H. Schoemaker, "How to Link Strategic Vision to Core Capabilities," *Sloan Management Review*, Fall 1992, p. 67.

[5]See, for example, Pierre Wack, "Scenarios: Shooting the Rapids," *Harvard Business Review*, November/December 1985, p. 146.

tions decisionmakers can take directly—as well as indirectly, through their influence on others.

A good defense planner will also identify a system of strategic indicators to trigger adaptive strategies, along with near-term courses of action decisionmakers may need to pursue to prepare or develop the strategies. Simply stated, the process is to develop strategies, test them, file them until needed, but exercise them periodically.[6]

In formal terms, collaboration is interaction among people that is intended to "create a shared understanding that none had previously possessed or could have come to on their own."[7] Collaboration technologies promise to improve the ability to coordinate action, share information, and understand information in order to facilitate inter- and intra-organizational teams. The following list shows three levels of collaboration; experienced defense analysts must be good at all of them.[8]

- *Level 1: Individual.* Individuals operating independently interact to selectively accommodate their own specific needs.

- *Level 2: Community of interest.* Groups of individuals exchange information in a shared community but not to achieve a common goal.

- *Level 3: Collaboration.* Individuals operate as a team to achieve a common goal by working together, sharing information, and thereby gaining new insights.

"Groupware" is software that supports the third level of collaboration. At its best, it applies the scientific method to the process of how groups use or should use analysis in making decisions. Analysis, in turn, can be considered good when it is a structured, systematic,

---

[6]On broad-based assumption-based planning and other styles of strategic planning, see Paul K. Davis and Zalmay Khalilzad, *A Composite Approach to Air Force Planning,* MR-787-AF, RAND, 1996.

[7]Michael Schrage, *No More Teams! Mastering the Dynamics of Creative Collaboration* (Doubleday, 1995), as quoted in P. A. Dargan, "The Ideal Collaboration Environment," April 2001 (available at http://www.stsc.hill.af.mil/CrossTalk/2001/apr/dargan.asp).

[8]Michael Schrage, *Shared Minds—The New Technologies of Collaboration,* Random-House, Inc., New York, 1990, as cited in P. A. Dargan.

traceable process of providing useful information to planners. This means that all data, inputs, assumptions, and methodologies are made transparent, and alternative decision paths are examined using logic chains to evaluate each one's advantages and disadvantages.

The first chapter in this part of the book is Chapter Nine, Paul Davis's "Exploratory Analysis and Implications for Modeling," in which he examines the consequences of uncertainty—not merely via standard sensitivity methods, but more comprehensively. Rather than going into excruciating detail on n^{th}-order effects, he uses a wide array of input variables (many well beyond what "experts" believe is plausible) to discover both the key uncertainties on which analysis may hinge and the primary drivers beside which all other variables pale in importance. This technique is useful primarily for studies in breadth (rather than depth), especially to gain a broad understanding of a problem area before dipping into details, or to see a forest rather than trees after detailed analysis. Hence, it is a good fit for capabilities-based planning. Davis describes techniques for doing exploratory analysis and explains how such analysis can be facilitated by multiresolution, multiperspective models (MRMPMs).

Chapter Ten, Dan Fox's "Using Exploratory Modeling," focuses on the practical issues associated with harnessing combat modeling and modern computers to explore a wide range of outcomes in order to understand the risks of engagement. In effect, he designs an experiment by identifying and then systematically varying experimental decision and risk variables to produce a range of outcomes. *Decision variables* represent policy alternatives (e.g., different levels of committed forces or alternative concepts of operations); *risk variables* (e.g., warning time) are given, not chosen. Measures of outcome may be simple (e.g., maximum kilometers that enemy forces advance) or complex (e.g., ratio of friendly to enemy losses). Combat simulation is used to create a matrix of results representing the outcomes for every combination of values for the experimental variables. Fox includes a comprehensive example that goes from designing the experiment to interpreting the results.

Stuart Starr's "Assessing Military Information Systems" is Chapter Eleven. Twenty-five years ago, Starr observes, several defense intellectuals sought to construct an ability to assess how much information systems contribute to military mission effectiveness. In line with this goal, substantial progress has been made in four areas: culture, process, tools, and experiments. Many of the principles and much of the guidance that have emerged in these areas are summarized in the revised 2002 version of *NATO Code of Best Practice for Command and Control Assessment.*[9] However, the changing geopolitical landscape poses daunting challenges for the assessment community. The contribution and impact of information systems must now be assessed in the context of not only emerging missions and complex, multidimensional information infrastructures ("infospheres"), but also the broader transformation of DoD itself.

Chapter Twelve is David Mussington's "The Day After Methodology and National Security Analysis." Mussington outlines the "Day After," an innovative gaming technique, developed by RAND, that examines strategic issues by playing out a scenario and then working backward to see how better decisionmaking could have improved the outcome. For this technique to work well, the scenario must be carefully designed and the testing process must be lengthy. Mussington describes how, in two concrete examples—one dealing with strategic information warfare, and one with the use of e-commerce technologies for money laundering—this approach illuminated strategy and policy questions. The most important issue in using this approach is how to remove the biases of the exercise designer or research sponsor from the scenario design or question formulation and still retain the policy relevance of the deliberations and findings. A readily usable *process* for evaluating scenario details and issue treatment is applied as an integral part of exercise development.

In the final chapter, "Using Electronic Meeting Systems to Aid Defense Decisions," Stuart Johnson explores another set of tools for improving the quality of defense deliberations. The basic problem being addressed is that, faced with complexity and uncertainty, individual planners risk becoming comfortable with familiar illusions. Moreover, there are limits to how well any one planner can imagine the future; and when planners work collectively, groupthink becomes a real risk. Johnson discusses a case in which an electronic meeting system (EMS) was used to help the Navy prepare for the 1997 Quadrennial Defense Review (QDR). Experts were first asked to

---

[9]Available at http://www.dodccrp.org/nato_supplnato.htm.

rank missions in terms of priority and likelihood (via anonymous voting, which was informed, after the fact, by group discussion). They were then asked to rank capabilities in terms of their contribution to each mission, and systems in terms of their contribution to each capability. This process resulted in a conclusion that probably would not have been recognized up front: command-and-control systems (C2) were of especial importance to naval operations and thus deserved to have their budget fenced off during the QDR process.

Chapter Nine

EXPLORATORY ANALYSIS AND IMPLICATIONS FOR MODELING

Paul K. Davis

The theme that runs through this book is that real-world strategy problems are typically beset with enormous uncertainties that should be central in assessing alternative courses of action. In the past, one excuse for downplaying uncertainty—perhaps treating it only through marginal sensitivity analysis around some "best-estimate" baseline of dubious validity—was the sheer difficulty of doing better. If analysis depended on the time it took to set up and run computer programs, then extensive uncertainty work was often ruled out. Today, however, extensive uncertainty analysis can be done with personal computers. Better software tools are needed, but existing commercial products are already powerful. There is no excuse for not doing better.

A key to treating uncertainty well is *exploratory analysis.* Its objectives are to (1) understand the implications of uncertainty for the problem at hand and (2) inform the choice of strategy and subsequent modifications. In particular, *exploratory analysis can help identify strategies that are flexible, adaptive, and robust.*[1] This chapter describes exploratory analysis, puts it in context, discusses enabling technology and theory, points to papers applying the ideas, and concludes with some challenges for those building models or developing enabling technology for modeling and simulation.

---

[1] Paul K. Davis, *Analytic Architecture for Capabilities-Based Planning, Mission-System Analysis, and Transformation*, MR-1513-OSD, RAND, 2002.

EXPLORATORY ANALYSIS

Definition

Exploratory analysis examines the consequences of uncertainty. In a sense, it is sensitivity analysis done right. Yet because it is so different in practice from what most people think of as sensitivity analysis, it deserves a separate name. It is closely related to scenario space analysis,[2] which dates back to 1983,[3] and to "exploratory modeling."[4] It is particularly useful for gaining a broad understanding of a problem domain before dipping into details. That, in turn, can greatly assist in the development and choice of strategies. It can also enhance "capabilities-based planning" by clarifying *when* (e.g., in what circumstances and with what assumptions about other factors) a given capability (e.g., an improved weapons system or enhanced command and control [C2]) will likely be sufficient or effective. This contrasts sharply with establishing a base-case scenario and an organizationally blessed model and database, and then asking, "How does the outcome for this scenario change if I have more of this capability?"

Exploratory analysis is an exciting development with a long history, including work in the 1980s and 1990s with RAND's RSAS (RAND Strategy Assessment System) and JICM (Joint Integrated Contingency Model). It is, however, only one part of a sound approach to analysis generally—a point worth pausing to emphasize.

Figure 9.1 shows how different types of models and simulations (including human games) have distinct virtues. The figure is specialized to military applications, but a more generic version applies broadly to a wide class of analysis problems. White rectangles indicate

---

[2]Paul K. Davis (ed.), *New Challenges for Defense Planning: Rethinking How Much Is Enough*, MR-400-RC, RAND, 1994.

[3]Paul K. Davis and James A. Winnefeld, *The RAND Strategic Assessment Center: An Overview and Interim Conclusions About Utility and Development Options*, R-2945-DNA, RAND, 1983.

[4]Stephen C. Bankes, "Exploratory Modeling for Policy Analysis," *Operations Research*, Vol. 41, No. 3, 1993; and Robert Lempert, Michael E. Schlesinger, and Steven C. Bankes, "When We Don't Know the Costs or the Benefits: Adaptive Strategies for Abating Climate Change," *Climatic Change*, Vol. 33, No. 2, 1996.

"good"—i.e., if a cell of the matrix is white, the type of model indicated in the left column is very effective with respect to the attribute indicated in the cell's column. In particular, analytic models (top left corner), which have low resolution, can be especially powerful with respect to their analytic agility and breadth. In contrast, they are very poor (have dark cells) with respect to recognizing or dealing with the richness of underlying phenomena, or with the consequences of both human decisions and behavior. In contrast, field experiments often have very high resolution (they may even be using the real equipment and people) and may be good or very good for revealing phenomena and reflecting human issues. They are, however, unwieldy and inappropriate for studying issues in breadth. The value of the type model for the particular purpose can often be enhanced a notch or two if the models include sensible decision algorithms or knowledge-based models that might be in the form of expert systems or artificial-intelligence agents.

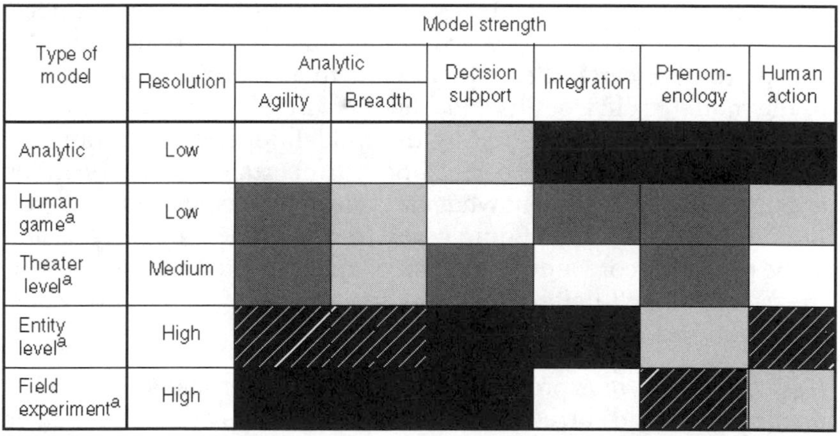

Type of model	Model strength						
	Resolution	Analytic		Decision support	Integration	Phenom-enology	Human action
		Agility	Breadth				
Analytic	Low						
Human game[a]	Low						
Theater level[a]	Medium						
Entity level[a]	High						
Field experiment[a]	High						

[a]Simulations.

NOTE: Assessments depend on many unspecified details. For example, agent-based modeling can raise effectiveness of most models, and small field experiments can be quite agile.

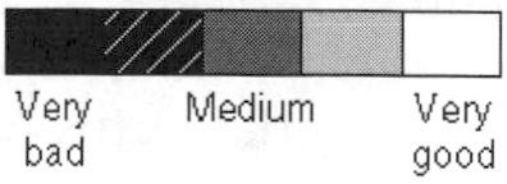

Very bad   Medium   Very good

Figure 9.1—Virtues of a Family of Models (Including Human Games)

Figure 9.1 is an exhortation to the Department of Defense (DoD) regarding the need to have *families of models and families of analysis.*[5] Unfortunately, it is usual for government agencies to depend more or less exclusively on a single model, which is a serious shortcoming.

Figure 9.1 reveals a niche for exploratory analysis: the top left-hand corner of the matrix, which emphasizes analytic agility and breadth of analysis. However, the technique can be used hierarchically with a suitably modularized system model. That is, one can do top-level exploration first, and then zoom in to explore in more detail—but using the same techniques—the consequences of various details within particular modules. This is easier said than done, however, especially with traditional models. Specially designed models make things much easier, as discussed in what follows.

Types of Uncertainty in Modeling

Exploratory analysis is about addressing uncertainty, but uncertainty comes in many forms. Parametric uncertainties arise from a model's inputs, from not knowing their precise values. They are not the same as structural uncertainties, which relate to questions about the form of the model itself: Does it reflect all the variables on which the real-world phenomena described by the model depend? Is the analytic form of the dependencies correct? Some uncertainties may be inherent because they represent what are called stochastic processes—the randomness of nature.[6] Some come from fuzziness or imprecision; some reflect discord among experts. Some relate to knowledge about the values of well-defined parameters; others refer to future values not yet known.

It is convenient to express uncertainties parametrically. Even when unsure about the correct form of the model, one can reflect uncertainty to some extent by having parameters that affect that form. For example, parameters may control the relative size of quadratic and exponential terms in an otherwise linear model. Or a discrete pa-

---

[5]Paul K. Davis, James H. Bigelow, and Jimmie McEver, *Analytical Methods for Studies and Experiments on "Transforming the Force,"* DB-278-OSD, RAND, 1999.

[6]The behavior of stochastic systems has a random component. Such systems are described with probabilistic equations.

rameter that is essentially a switch might determine which of a set of distinct analytic forms applies. Some parameters may apply to the fixed aspect of a model; others may apply to a random aspect. In taking this approach, one needs to keep straight how the different uncertainties come into play.[7]

Types of Exploratory Analysis

Ways to Conduct Exploratory Analysis.[8] One form of exploratory analysis is input, or parametric, exploration, which involves running models across the space of cases defined by plausible discrete values of the parameters. This is done not one at a time, as in normal sensitivity analysis, and not around some presumed base-case set of values, but, rather, for all the combinations of values defined by an experimental design. The results, which may number from dozens to hundreds of thousands or more, can be explored interactively with modern displays. Within perhaps one half-hour, a good analyst can often gain numerous important insights that were previously buried. He can understand not just which variables "matter," but when they matter. For example, he may find that outcomes are insensitive to a given parameter for the so-called base case but are quite sensitive for other plausible assumptions. That is, he may identify *when* the parameter is important. For capabilities-based planning for complex systems, this can be distinctly valuable.[9]

---

[7]See the appendix for an example that uses Lanchester equations, chosen for familiarity rather than for current usefulness.

[8]See Paul K. Davis, David C. Gompert, and Richard L. Kugler, *Adaptiveness in National Defense: The Basis of a New Framework*, IP-155, RAND, 1996; Bankes, "Exploratory Modeling"; Arthur Brooks, Steve C. Bankes, and Bart Bennett, *Weapon Mix and Exploratory Analysis: A Case Study*, DB-216/2-AF, RAND,1997; Lempert, Schlesinger, and Bankes, "When We Don't Know"; National Research Council, *Modeling and Simulation*, Vol. 9, *Technology for the United States Navy and Marine Corps, 2000–2035*, National Academy Press, Washington, DC, 1997; and Granger Morgan and Max Henrion, *Uncertainty: A Guide to Dealing with Uncertainty in Quantitative Risk and Policy Analysis*, Cambridge University Press, Cambridge, MA, 1992 (reprinted in 1998). The book by Morgan and Henrion is an excellent treatment of uncertainty in policy analysis generally.

[9]Many examples in military analysis involve warning time. Some capabilities, such as those of forward-deployed systems, are especially important only when warning time is short. If standard scenarios assume considerable warning time, such capabilities can be undervalued.

A complement to parametric exploration is probabilistic exploration, in which uncertainty about input parameters is reflected by distribution functions representing the totality of one's so-called objective and subjective knowledge. Using analytic or Monte Carlo methods, the resulting distribution of outcomes can be calculated. This can quickly give a sense for whether—all things considered—uncertainty is particularly important. In contrast to displays of parametric exploration, the output of probabilistic exploration gives little visual weight to improbable cases in which various inputs all have unlikely values simultaneously. Probabilistic exploration can be very useful for a condensed "net assessment." Note that this use of probability methods is different from using them to describe the consequences of a stochastic process within a given simulation run. Indeed, one should be cautious about using probabilistic exploration, because one can readily confuse variation across an ensemble of possible cases (e.g., different runs of a war simulation) with variation within a single case (e.g., fluctuation from day to day within a single simulated war). An unknown constant parameter for a given simulated war is no longer unknown once the simulation begins, and simulation agents representing commanders should perhaps observe and act upon the correct values within a few simulated days. Despite these subtleties, probabilistic exploration can be quite helpful.

After initial work with both parametric and probabilistic exploration, the preferred approach is *hybrid exploration*. It may be appropriate to parameterize a few key variables that are under one's control (purchases, allocation of resources, and so on) while treating the uncertainty of other variables through uncertainty distributions. Analysts might also want to parameterize a few of the principal variables characterizing the future context in which strategy must operate. In military affairs, one might parameterize assumed warning time and size of threat. There is no general procedure here; instead, the procedure should be suitable to the problem at hand. In any case, the result of such exploratory analysis can be a comprehensible summary of how known classes of uncertainty affect the problem at hand.

Consider the following examples of exploratory analysis. Figure 9.2 displays a mid-1990s parametric exploration of what is required militarily to defend Kuwait against a future Iraqi invasion by interdicting

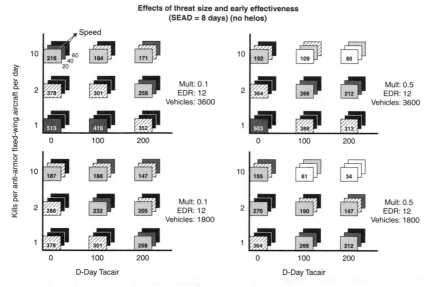

Figure 9.2—Data View Display of Parametric Exploration

the attacker's movement with aircraft and missiles.[10] Each square in the figure represents a particular model case (i.e., a specific choice of all the input values). Taken together, the four panels summarize parametric exploration in five variables (the x, y, and z axes of each panel, one for a row of two panels, and one for a column of two panels). With four such pages, one can cover results for seven variables. The outcome of a given simulation is represented by the pattern of a given square. A white square represents a good case, in which the attacker penetrates only a few tens of kilometers before being halted. A black square represents a bad case, in which the attacker penetrates deep into the region that contains critical oil facilities. The other patterns represent in-between cases. The number in each square gives the penetration distance in kilometers.

---

[10]Paul K. Davis and Manuel J. Carrillo, *Exploratory Analysis of "The Halt Problem": A Briefing on Methods and Initial Insights,* DB-232, RAND, 1997.

To generate such results for a sizable scenario space, RAND has often used a program called Data View.[11,12,13] After running the thousands or hundreds of thousands of cases corresponding to an experimental design for parametric exploration, one explores the outcome space by choosing interactively which of the parameters to vary along the x, y, and z axes. The remaining parameters then have values shown alongside the graph. These can be changed by clicking on one of them and selecting from the menu that comes up.

Figure 9.3 shows a screen image from some recent work with Analytica on the same interdiction-of-invader-forces problem treated in Figure 9.2. In this case, the graphical display of results is more traditional. Outcome is measured along the y axis rather than by a color or pattern, and one of the independent variables is plotted along the x axis. A second variable—D-Day (the day war commences) shooters—is reflected in the family of curves. The other independent variables appear in the rotation boxes at the top. As with Data View, one changes parameter values by selecting from a menu of values. Such interactive displays allow one to "fly through the outcome space" for many independent variables (parameters), in this case nine. More parameters could have been varied interactively, but the display was still quickly interactive for the given model and computer being used (a Macintosh PowerBook G3 with 256 MB of RAM).

---

[11]This was developed at RAND in the mid-1990s by Stephen Bankes and James Gillogly.

[12]Other personal-computer tools can be used for the same purpose and the state of the art for such work is advancing rapidly. A much improved version of Data View, called CARS[TM], is under development by Evolving Logic (www.evolvinglogic.com). For those who do their modeling with Microsoft EXCEL[TM], there are plug-in programs that provide statistical capabilities and some means for exploratory analysis. Two such tools are Crystal Ball[®] (www.decisioneering.com) and Risk[®] (www.palisade.com/html/risk.html). For a number of reasons, however (visual modeling, array mathematics, etc.), my colleagues and I have in recent times most often used the Analytica[®] modeling system. Analytica is an outgrowth of the Demos system developed by Max Henrion and Granger Morgan at Carnegie Mellon University; it is marketed by Lumina (www.lumina.com).

[13]For more recent exploratory analysis work, see Paul K. Davis, Jimmie McEver, and Barry Wilson, *Measuring Interdiction Capabilities in the Presence of Anti-Access Strategies: Exploratory Analysis to Inform Adaptive Strategy in the Persian Gulf*, MR-1471-AF, RAND, 2002.

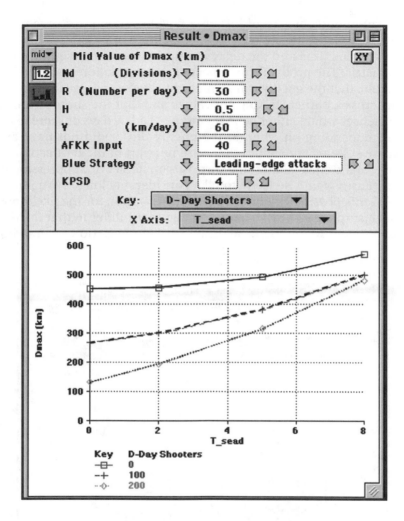

Figure 9.3—Analytica Display of Parametric Exploration
(Simultaneous Exploration of Nine Parameters)

So far, the examples have focused on parametric exploration. Figure 9.4 illustrates a hybrid exploration.[14] It shows the distribution of

---

[14]Paul K. Davis, David C. Gompert, Richard J. Hillestad, and Stuart Johnson, *Transforming the Force: Suggestions for DoD Strategy*, IP-179, RAND, 1998.

simulation outcomes resulting from having varied most parameter values "probabilistically" across an ensemble of possible wars, but with warning time and the delay in attacking armored columns left parametric. The probabilistic aspect of the calculation assumed, for example, that the enemy's movement rate had a triangular distribution across a particular range of values and that the suppression of air defenses would either be in the range of a few days or more like a week, depending on whether the enemy did or did not have air-defense systems and tactics that were not part of the best estimate. That is, if the enemy had some surprises up its sleeve, suppression of air defenses would be likely to take considerably longer. We represented this possibility with a discrete distribution for the likelihood of such surprises. The two curves in Figure 9.4 differ in that the one with crosses for markers assumes that interdiction of moving

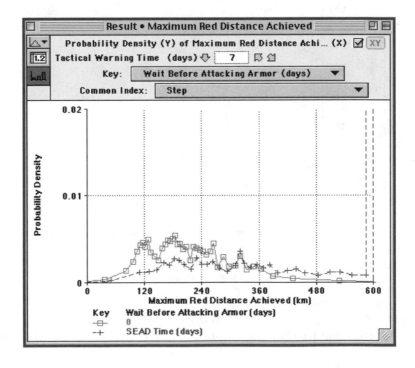

Figure 9.4—Display of "Probabilistic" Exploration

columns waits for suppression of enemy air defenses (SEAD). The other curve assumes that interdiction begins immediately because the aircraft are assumed stealthy or the defenses nonexistent.

This depiction of the problem shows in one display how widely outcomes can vary and how outcome distribution can be complex. The non-stealthy-aircraft case shows a considerable spike at the right end, where many cases pile up because, in the simulation, the attacker halts once he has reached an objective at about 600 km. Note that the mean is not a good metric: the variance is huge, and the outcome may be bimodal or even multimodal.

Advanced Concepts. The results just discussed are from analyses accomplished in recent years for DoD. Looking to the future, much more is possible with computational tools. Much better displays are possible, and, even more exciting, computational tools can be used to aid in the search process of exploration. For example, instead of "clicking" through the regions of the outcome space, tools could automatically locate portions of the space in which particular outcomes are found. Insights could be fine tuned by clicking around in that much more limited region of the outcome space. Or, if the model is itself driven by the exploration apparatus, the apparatus could search for outcomes of interest and then focus exploration on those regions of the input space. That is, the experimental design could be an output of the search rather than an input of the analysis process.

ENABLING EXPLORATORY ANALYSIS

In principle, exploratory analysis can be accomplished with any model. In practice, it becomes difficult with large and complex models. If F represents the model, it can be considered to be simply a complicated function of many variables. How can we run a computerized version of F to understand its character? If F has M inputs with uncertain values, then we could consider N values for each input, construct a full-factorial design or some properly sampled subset, run the cases, and thereby have a characterization. However, the number of such cases would grow as NM for full-factorial analysis, which quickly gets out of hand even with big computers. Quite aside from issues of setup and run time, comprehending and communicating the consequences becomes very difficult if M is large. Suppose someone asked, "Under what conditions is F less than danger_point?" Given sufficiently powerful computers and enough time, one could create a database of all the cases, after which one could respond to the question by spewing out lists of the cases in which F fell below danger_point. The list, however, might go on for many pages, perhaps even thousands. What would be done with the list? This is one manifestation of what might be called the curse of dimensionality.

It follows that—even with a perfect high-resolution model and incredibly speedy computers—abstractions, such as aggregate representations, will still be necessary. In the most usual cases, in which the high-resolution model is by no means perfect, abstractions allow analysts to ponder the phenomena in meaningful ways, with relatively small numbers of cognitive chunks to deal with. People can reason with 3, 5, or, perhaps, even 10 such cognitive chunks at a time, but not with hundreds. If the problem is truly complex, ways must be found to organize the reasoning—i.e., the problem must be decomposed. One ends up using principles of modularity and hierarchy. To head off a rejoinder commonly made at this point by aficionados of networking technology, the need for an aspect of hierarchical organization is inescapable in most systems of interest, even if the system is highly distributed and relatively nonhierarchical. Everyone can observe this when interacting with the World Wide Web.

A corollary of the need for abstractions is the need for models that use the various abstractions as inputs. It is not sufficient to display the abstractions as intermediate outputs (displays) of the ultimate detailed model. The reasons include the fact that when a decisionmaker asks a what-if question using abstractions, there is a 1:n mapping problem in translating his question into the inputs of a more detailed model. That is, the decisonmaker asks, "What if we had 25 percent more capability?" but the detailed model represents many capabilities. What assumptions about these many capabilities would correspond best to the decisionmaker's question? In contrast, a more aggregated model may already have the concept of overall capability; it can then address the decisionmaker's question directly. That is, it accepts the decisionmaker's abstractions as inputs.

Given the need for abstractions, how do we find them and how do we exploit them? The approaches fall into two groups, one for new models and the other for existing, or legacy, models.

With new models, the issue is how to design, and the options of interest are

- Design the models and model families top down so that significant abstractions are built in from the start, but do so with enough understanding of the microscopics that the top-down design is valid.[15]

- Design the models and model families bottom up, but with enough top-down insight to build in good intermediate-level abstractions from the start.[16]

- Do either or both of the above, but with designs taken from different user or theoretical perspectives.

Note that this list does not include a pure top-down or bottom-up design approach. Only seldom will either generate a good design of a complex system. Note also the idea of alternative perspectives. This recognizes that many abstractions are not unique; to the contrary, there are different ways of viewing what the key factors of the problem really are (e.g., those in combat arms typically conceive military problems differently than logisticians do).

Only sometimes is there the opportunity to design from scratch. More typically, existing models must be used (or adapted and used). Moreover, the model "families" available are often families more on the basis of assertion or hope than lineage. What does one then do? Some possibilities are as follows:

- Given existing models developed at high levels of resolution, study the model and the questions that users ask of the model to discover useful abstractions. For example, one may discover that inputs X, Y, and Z only enter the computations as the product

---

[15]Paul K. Davis and James H. Bigelow, *Experiments in Multiresolution Modeling (MRM)*, MR-1004-DARPA, RAND, 1998.

[16]Paul K. Davis and James H. Bigelow, *Motivated Metamodels: Synthesis of Cause-Effect Reasoning and Statistical Metamodeling*, MR-1570-AF, RAND, 2003.

XYZ. If so, then XYZ may be a natural abstraction. Or, perhaps decisionmakers ask questions in terms of concepts such as force strength or force ratio, indicating that these are significant abstractions. For mature models, the obvious place to look is the list of displays that have been added over time to provide views into the internal workings of the model.

- Apply statistical machinery to search for useful abstractions. For example, if X, Y, and Z are inputs, such machinery might test to see whether the system's behavior correlates not just with X, Y, and Z, but with XY, XZ, YZ, and XYZ. If the computation does, in fact, depend only on XYZ, that fact will show up from the statistical analysis.

- Idealize the system and develop a "formal" mathematical representation (formal in the sense of being expressed symbolically without necessarily having the intention of computing the various terms and factors) that provides hints about the model's likely behavior. For example, such a representation might be much too complex to "solve," but, if coupled with some physical reasoning and a search for postulated simplifications, it might highlight the likelihood of an overall exponential decay, or an inverse dependence on one input, or various other nonlinearities that otherwise one might think to test for. It might suggest natural *aggregation fragments*, such as the product XYZ mentioned above. In practice, this approach is most powerful if one considers the problem from different perspectives that suggest different but plausible simplifications.[17,18]

This list is less straightforward than it first appears. The first approach is perhaps a natural activity for a smart modeler/programmer who begins to study an existing program—if, and only if, he is also a believer in higher-level depictions of the problem. The second approach seems to be favored by mathematically oriented individuals who lack enough class knowledge to take the first approach, or who

---

[17]One example of a simplification is the assumption that an integral is perhaps approximately equal to a representative value of the integrand times the effective width of the integration interval, and that this width is proportional to something physically straightforward.

[18]Davis and Bigelow, *Motivated Metamodels.*

believe—based sometimes on disciplinary faith—that such statistical procedures will prove successful and that those looking for more phenomenological abstractions are fooling themselves. The third approach is a hybrid. It argues that one should use one's understanding of phenomenology, and theories of system behavior, to gain insights about the likely or possible abstractions. Only then should one crank the statistical machinery. Where it is feasible, this is the stronger approach.

Using Occam's Razor

An interesting tension arises in discussing how to form suitable abstractions. The principle of Occam's razor requires that one prefer the simplest explanation and, thus, the simplest model. Some, particularly enthusiasts of statistical approaches, tend to interpret this principle to mean that one should minimize the number of variables in a model. They tend to focus on data (natural or simulation generated) and to avoid adding variables for the purpose of "explanation" or "phenomenology" if the variables are not needed to predict the data. Instead, they prefer to "let the data speak." In contrast, subject-area phenomenologists may prefer to enrich the depiction by adding variables that provide a better picture of cause-effect chains. This, however, goes well beyond what can be supported with meager experimental data.

To not violate the Occam's Razor principle, one must remember the principle's longer form: Adopt the simplest explanation that truly explains all there is to explain—but nothing simpler! This should include phenomena that one "knows about," even if they are not clearly visible in the limited data (e.g., historical data on who won various battles with what overall attrition). I would add to this the old admonition (perhaps made first by Massachusetts Institute of Technology's Jay Forrester) to remember that to omit showing a variable one knows about may be equivalent to assuming its value to be 1 (as a multiplier) or 0.

It is sometimes useful to have a competition among approaches. For example, phenomenologists working a problem may be utterly convinced that it must be described with complex computer programs having hundreds or thousands of data elements. In such a case, it may be useful to study output behavior with a metamodel (also called a repro model and response surface).[19] In one instance with which I am familiar, such work by colleague James Bigelow showed that despite many man-years of effort building and feeding a complex ground-force model, results were strongly dominated by a single higher-level abstraction: theater-level force ratio. The implication was not that real combat is dominated only by theater-level force ratio, but, rather, that various assumptions and compromises made in developing the detailed model undercut any claims that its greater complexity and expense were adding predictive value relative to simpler models.

Although the discussion above distinguishes sharply between the case of new models and old ones, the two are connected. In essence, working with existing models should often involve sketching what the models *should* be like and how models with different resolution *should* connect substantively. That is, working with existing models may mean having to go back to design issues. If this seems suspicious, ask yourself how often you have found it easier to rederive a model and then decipher a program you have been given than to wade through the program on its own terms.

Multiresolution, Multiperspective Modeling and Model Families

Abstractions, usually aggregations, are fundamental in exploratory analysis. Finding suitable abstractions, relating them, and conducting both high-level exploratory analysis and appropriate zooming into detail are greatly facilitated if models are designed in a special way. This is the subject of multiresolution, multiperspective modeling (MRMPM). Although the subject relates most directly to new models, it is also relevant to working with legacy models in preparing for exploratory analysis.

Multiresolution modeling (MRM) is building a single model, a family of models, or both to describe the same phenomena at different levels of resolution and to allow users to input parameters at those dif-

---

[19]A metamodel is a simple model that reproduces the aggregate behavior of a more complex model, as judged by statistical comparisons over many cases.

ferent levels depending on their needs.[20] Variables at level n are abstractions of variables at level n+1. MRM has also been called variable- or selectable-resolution modeling.[21] Figure 9.5 illustrates MRM schematically. It indicates that a higher-level model (model A) itself has more than one level of resolution. It can be used with either two or four inputs. However, in addition to its own MRM features, it has input variables that can be specified directly or determined from the outputs of separate higher-resolution models (models B and C,

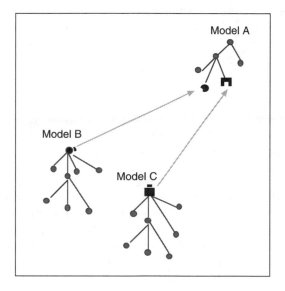

Figure 9.5—Multiresolution Family of Models

---

[20]Davis and Bigelow, *Experiments in Multiresolution Modeling.*

[21]Unfortunately, some authors use the term *multiresolution models* to mean only that there are *outputs* at different levels of detail or that a model happens to treat different phenomena asymmetrically (e.g., with detail for combat processes and a more aggregate depiction for logistics). Still other authors claim multiresolution features for models because their objects are hierarchical. The essence of multilevel resolution, as I use the term here (and as discussed in National Research Council, *Modeling and Simulation*), is having multiple levels of abstraction for key *processes,* such as attrition, movement, and communications. Achieving such MRM is quite challenging. RAND work on the subject dates back to the late 1980s.

shown as "on the side," for use when needed). In principle, one could attach models B and C in the software itself, creating a bigger model. However, in practice there are tradeoffs between doing this and keeping the more detailed models separate. For larger models and simulations, a combination single-model/family-of-models approach is desirable because it balances the needs for analytic agility and complexity management.

MRM is not enough, however, because, as noted earlier, different applications require different abstractions even if the resolution is the same—i.e., different "perspectives" are legitimate and important. Perspectives are distinguished by the conception of the system and the choice of variables; they are analogous to alternative "representations" in physics or engineering. Designing for both multiple resolution and multiple perspectives can be called MRMPM.[22]

With MRMPM models (single models or families), the concepts and predictions among levels and perspectives have to be connected (and reconciled). It is often assumed that the correct way to do this is to calibrate upward: treat the information of the most detailed model as correct and use it to calibrate the higher-level models. This, indeed, is often appropriate. The fact is, however, that the more detailed models almost always have omissions and shortcomings.[23] Models at higher levels, and from different perspectives, address some of them explicitly. Further, the different models of a family draw on different sources of information—ranging from doctrine or even "lore" on one extreme to physical measurements on a test range at the other. One class of information is not inherently better than another; it is simply different.

Figure 9.6 makes the point that members of a multiresolution model family should be *mutually* calibrated, with information flows in both directions. In the military domain, for example, low-resolution historical attrition or movement rates may be used to help calibrate more-detailed models predicting attrition and movement. This is not

---

[22]Paul K. Davis, "Exploratory Analysis Enabled by Multiresolution, Multiperspective Modeling," in Jeffrey A. Joines et al. (eds.), *Proceedings of the 2000 Winter Simulation Conference*, 2000.

[23]For example, detailed models often have a rich depiction of physical considerations but only a minimal representation of behaviors that adapt to circumstances.

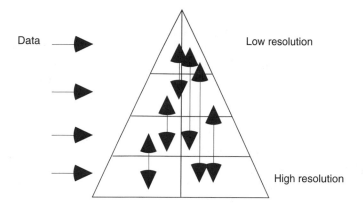

Figure 9.6—Mutual Calibration of a Family of Models

straightforward, because of the 1:n mappings. It is often done crudely, by applying an overall scaling factor (fudge factor) rather than correcting the more atomic features of the model, but it is likely to be something with which readers are familiar. However, much calibration is indeed upward. In a study using detailed order-of-battle information, for example, inputs on the number of "equivalent divisions" or "equivalent F-15 aircraft" used in abstract models can be computed from the data feeding high-resolution models. Furthermore, at least in principle, the attrition coefficients' dependence on situation (e.g., open versus wooded terrain for ground forces) should be informed by high-resolution work.

So, given their desirability, how should families of models be built? Or, given preexisting models, how "should" they relate before they are connected as software or used for mutual calibration? The first design principle may be to recognize that there are limits to what is feasible. In particular, there are limits to how well low-resolution models can be consistent with high-resolution models. *Approximation is a central concept from the outset.* Several points are especially important in thinking about this:[24]

---

[24]Davis and Bigelow, *Experiments in Multiresolution Modeling.*

- Consistency between two models of differing resolution should be assessed in the context of how the models are being used. What matters is not whether they generate the same final state of the overall system, but whether they generate approximately the same results in the application. That may be something as specific as summary graphs or rank ordering of alternatives. Another way to put this is to be practical, not theological, about how much detail is needed.

- The implications for consistency of aggregation and disaggregation processes must also be judged in context. Some disaggregation assumptions represent aggregate-level knowledge not necessarily reflected in the most detailed model.

- Comprehensive MRM is very difficult or impossible for complex modeling and simulation,[25] but having even some MRM can be far more useful than having none at all. MRM is not an all-or-nothing matter.

- The various members of an MRM family will typically be valid for only portions of the system's state space. As one moves from one region to another, valid description may require that parameter values or even the structure of the model itself be changed.

- Mechanisms are therefore needed to recognize different situations and shift models. In simulations, human intervention is one mechanism and agent-based modeling is another.[26]

- Valid MRM will often require stochastic variables represented by probability distributions, not merely mean values. Further, valid aggregate models must sometimes reflect correlations among variables that might naively be seen as probabilistically independent.

---

[25]For an excellent theoretical discussion of this, see Robert Axtell, *Theory of Model Aggregation for Dynamical Systems with Applications to Problems of Global Change*, Ph.D. dissertation, Carnegie-Mellon University (available from University Microfilms International, Ann Arbor, MI). Axtell's discussion, however, fails to emphasize the key significance of approximations. As a result, it is more pessimistic than my own work.

[26]Agent-based models include modules that represent (i.e., serve as agents for) adaptive entities, such as individual humans or groups. The basic ideas are discussed in most books dealing with "complex adaptive systems."

With these observations up front, the ideal for MRM is a hierarchical design for each MRM process, as indicated earlier, in Figure 9.5.

Models and analysis methods for exploratory analysis should have a number of characteristics. First, they should be able to reflect hierarchical decomposition through multiple levels of resolution and from alternative perspectives representing different "aspects" of a system. For example, one model might decompose a system into its organizational components, another might focus on different component processes, and yet another might follow component functions.

Less obviously, models and analysis methods should also include realistic mechanisms describing how the natural entities of the system act, react, adapt, mutate, and change. These mechanisms should reflect the relative "fitness" of the original and emerging entities for the environment in which they are operating. Many techniques are applicable here, including game theory methods and others that may be relatively familiar to readers. However, the most fruitful new approaches are those typically associated with the term *agent-based modeling*. These include submodels that act "as the agents for" specific entities—say, political leaders and military commanders or (at the other extreme) infantry privates on the battlefield. In practice, such models need not be exotic: they may correspond to some relatively simple heuristic, or intuitive, decision rules or to some well-known (though perhaps complex) operations-research algorithm. But to have such decision models is quite different from depending on scripts.

Because it is implausible that closed computer models will be able to meet the above challenge in the foreseeable future, the family of "models" should allow for human interaction—whether in human-only seminar games, small-scale model-supported human gaming, or distributed interactive simulation. This runs against the grain of much common practice, which imputes too much virtue to "closed models" that generate readily reproducible results.

The last item in the bulleted list above is often ignored in today's day-to-day work, even by good analysts who have a family of models. Often, when they seek to use models at different levels of resolution analytically, they decide on a highest-level model to be used for excursions—i.e., for examining sensitivities. They then "calibrate" this highest-level model by using one or more detailed models. For example, they might use the Brawler model of air-to-air combat between small groups of aircraft in different groupings; they would then use results of that work to calibrate the air-to-air model of a theater-level depiction, such as in the TACWAR, JICM, Thunder, or START models. This is not easy. However, the analysts would sit down, talk, draw sketches, and so on, until they gained a sense of how to go about the calibration. Ultimately, for a particular study done on a limited budget and time scale (as most are) they might use expected-value outcomes of "representative" air-to-air engagements in Brawler to set attrition coefficients in the theater-level model. This might or might not be "correct," because the relationship between the engagement level and theater level is very complex: in a real air war, there may be thousands of engagements with a wide variety of characteristics, and how to aggregate is not so clear. For example, one might imagine that 80 percent of engagements are "normal" but have little effect on relative force levels, while 20 percent of engagements are of a different character and lead to one side annihilating the other's aircraft with no losses of its own. The overall time dependence of relative force levels, then, might be dictated by the unusual, nonrepresentative engagements. However, focusing on these unusual cases in doing the calibration might outrageously exaggerate one or both of the attrition rates. The "correct" way to go about the calibration would necessarily involve explicit, study-dependent integration over classes of engagement.

Sometimes, the higher-level model inputs need to be stochastic. Figure 9.7 illustrates the concept schematically for a simple problem. Suppose that a process—for instance, one computing the losses to aircraft in air-to-air encounters—depends on five inputs: Q, X, Y, a, and b. But suppose that the outcome of ultimate interest involves many instances of that process with different values of X and Y (e.g., different per-engagement numbers of Red and Blue aircraft). An abstraction of the model might depend only on Q, a, and b (e.g., overall attrition might depend on only numbers of Red and Blue aircraft, their relative quality, and some C2 factor). If the abstraction shown is to be valid, the variable Z should be consistent with the higher-resolution results. However, if it does not depend explicitly on X and Y, then there are "hidden variables" in the problem, and Z may

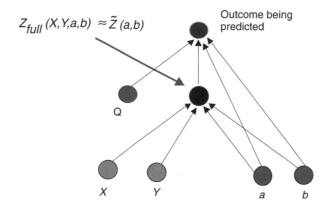

Figure 9.7—Input to Higher-Level Model May Be Stochastic

appear to be a random variable, in which case the predicted outcome would be a random variable. This randomness might be ignored if the distribution were narrow enough, but that might not be the case.

To compute what "should be," one would relate the probability density to a constrained integral over X and Y, appropriately weighting on the basis of likelihood and restricting the integration to regions where Z has the value of interest.

In the past, such calibrations have been rare, in significant part because analysts have lacked both theory and tools for doing things better. The "theory" part includes not having good descriptions of how the detailed model should relate to the simplified one. The tool part includes not being able to define the set of runs that should be done (representing the integral of Figure 9.7) and then to actually make those runs.

Ideally, such a calibration would be dynamic within a simulation. Moreover, it would be easy to adjust the calibration to represent different assumptions about command, control, communications, computers, intelligence, surveillance, and reconnaissance (C4ISR), as well as about tactics. We cannot do such things today, because modeling technology and practice are not up to it yet.

LESSONS FROM RECENT EXPERIENCE

Both exploratory analysis and MRM/MRMPM are relatively new concepts, but there is a growing body of examples to illustrate their practicality for addressing problems—for instance, the problem of halting an invading army using precision fires from aircraft and missiles.[27] The most recent aspects of that work included understanding in some detail how the effectiveness of such fires are affected by details of terrain, enemy maneuver tactics, certain aspects of command and control, and so on. This provided a good test bed for exploring numerous aspects of MRMPM theory.

We developed a multiresolution personal-computer model (PEM),[28] written in Analytica, to understand and extend to other circumstances the findings from entity-level simulation of ground maneuver and long-range precision fires. A major part of this work was learning how to inform and calibrate PEM to the entity-level work. There was no possibility, in this instance, of revising the entity-level model. Nor, in practice, did we have a good enough understanding of the model to construct a comprehensive calibration theory. Instead, we had to construct a new, more abstract model and attempt to impose some of its abstractions on the data from runs of the entity-level simulation in prior work, plus some special runs made for our purposes. Had we had the intermediate-level PEM several years earlier, we could have used it both to define adaptations of the entity-level model that would have generated some of the abstractions we needed and to better define the experiments conducted with the high-resolution model. Instead, we had to make do with the situation

---

[27]Our work on precision fires is discussed in Davis, Bigelow, and McEver, *Analytical Methods for Studies and Experiments*; Jimmie McEver, Paul K. Davis, and James H. Bigelow, *EXHALT: An Interdiction Model for Exploring Halt Capabilities in a Large Scenario Space*, MR-1137-OSD, RAND, 2000; Paul K. Davis, James H. Bigelow, and Jimmie McEver, *Effects of Terrain, Maneuver Tactics, and C4ISR on the Effectiveness of Long-Range Precision Fires: A Stochastic Multiresolution Model (PEM) Calibrated to High-Resolution Simulation*, MR-1138-OSD, RAND, 2000; and Davis, Bigelow, and McEver, *Effects of Terrain*. Some of this work was also used in the summer study of the Defense Science Board and is reflected in Eugene C. Gritton, Paul K. Davis, Randall Steeb, and John Matsumura, *Ground Forces for a Rapidly Employable Joint Task Force: First-Week Capabilities for Short-Warning Conflicts*, MR-1152-OSD, RAND, 2000.

[28]Davis, Bigelow, and McEver, *Effects of Terrain*.

we found ourselves in. The result is a case history with what are probably some generic lessons learned.

Figure 9.8 illustrates one aspect of our multiresolution PEM approach. The figure shows the data flow within a PEM module that generates the impact time (relative to the ideal impact time) for a salvo of precision weapons aimed at a packet of armored fighting vehicles observed by C4ISR assets at an earlier time. Other parts of the PEM combine information about packet location versus time and salvo effectiveness for targets that happen to be within the salvo's "footprint" at the time of impact in order to estimate the effectiveness of the precision weapons. For the salvo-impact-time module, Figure 9.8 shows how the PEM is designed to accept inputs as detailed as whether there is enroute retargeting of weapons, the C2 latency time, and weapon flight time. However, it can also accept more aggregate inputs, such as time from last update. If the input variable "resolution of last update calculation" is set "low," then time from last update is specified directly as input; if not, it is calculated from the lower-level inputs.

Being able to depict the problem as in Figure 9.8, and to provide users the option of what inputs to use, has proven very useful—both for analysis itself and for communicating insights to decisionmakers in communities ranging from the C4ISR community to the programming and analysis community. In particular, the work clarified how the technology-intensive work of the C4ISR acquisition community relates to higher-level strategy problems and analysis of such problems at the theater level.

Another companion piece describes how, in developing the PEM and a yet more abstract model (EXHALT) used for theater-level halt-problem analysis, we experimented with methods for dealing with the multiperspective problem.[29] Perhaps the key conclusion of this particular work is that MRMPM rather demands a building-block approach that emphasizes study-specific assembly of the precise model needed. Although we had some success in developing a closed

---

[29]Jimmie McEver, Paul K. Davis, and James H. Bigelow, "Implementing Multiresolution Models and Families of Models: From Entity Level Simulation to Personal-Computer Stochastic Models and Simple 'Repro Models,'" *SPIE 2000*, Orlando, FL, April 2000.

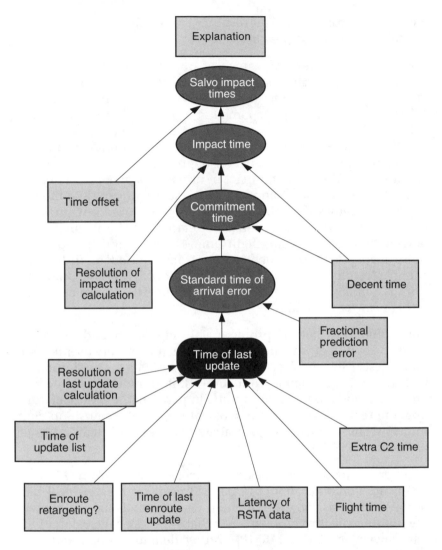

**Figure 9.8—A Multiresolution, Multiperspective Design
for Salvo-Impact-Time Module**

MRMPM model with alternative user modes representing different
demands for resolution and perspective (e.g., the switches in Figure
9.8), it proved impossible to do very much in that regard: the number

of interesting user modes and resolution combinations simply precludes being able to wire in all the relevant user modes. Moreover, the explosion of complexity occurs very quickly. Thus, despite the desire of many users to have a black-box machine that can handle all the cases and perspectives of interest, it seems a fundamental reality that at-the-time assembly from building blocks, not prior definition, is the stronger approach. This was as we expected, but even more so.

The ultimate reason for the building-block conclusion is that even in the relatively simple problem examined, the real variable trees (akin to data-flow diagrams) are bushy rather than rigorously hierarchical. Furthermore, the different legitimate perspectives can simply not all be accommodated simultaneously without making the code itself very complicated to follow. In contrast, we found it easy to construct the model needed quickly—in hours rather than days or weeks—as the result of our building-block approach, visual modeling, use of array mathematics, and strong, modular design.

As powerful as current personal computer tools are in comparison with those of past years, they are still not up to the challenge of making the building-block/assembly approach rigorous, understandable, controllable, and reproducible without unrealistically high levels of modeler/analyst discipline. Also, the search models for advanced exploratory analysis are not yet well developed. Thus, there are good challenges ahead, not just for the model builders and users, but also for the community that builds the enabling technology.

Appendix

REFLECTING UNCERTAINTY WITH PARAMETERS, AN EXAMPLE

As an example, consider a model that describes the rate at which Red and Blue suffer attrition in combat according to a Lanchester square law:

$$\frac{d\tilde{R}}{dt} = -\tilde{K}_b \tilde{B}(t) \quad \frac{d\tilde{B}}{dt} = -\tilde{K}_r \tilde{R}(t) ,$$

where the attrition coefficients for Red and Blue have both deterministic and stochastic parts, each of which is subject to uncertainty. The equation for Red says that the quantity of Red capability decreases in proportion to the quantity of Blue (because Blue is shooting at Red):

$$\tilde{K}_b(t) = K_{b0}\left[1 + c_b \tilde{N}_b(t; \mu, \sigma_b)\right] \quad K_{r0}\left[1 + c_r \tilde{N}_r(t; \mu, \sigma_r)\right] .$$

Here, K_{b0} and K_{r0} are average attrition rates for a given war. They may be highly uncertain (factors of 2, 3, or more), but they are constant within a war. That is, before the war, we may not know the sides' average effectivenesses, but they exist. This said, attrition will vary from battle to battle and from time period to time period within a given war. Such variation can be regarded as a stochastic process. These effects are reflected by the bracketed factors, above, where N_r and N_b are assumed to be normal random variables with means of m and standard deviations. Their parameters are also uncertain, perhaps strongly so, but it is a different kind of uncertainty than that about the average attrition for a given war.[30]

---

[30]I distinguish between deterministic uncertainty and stochastic processes, but both may be treated by the same mathematical tools, such as probability distributions. The distinction is important, however. For example, a commander discovering that his losses to attrition were three times what he expected on the first day of war—and ascribing that attrition to the unanticipated effectiveness of certain weapons—should not imagine that tomorrow is another day, that stochastic variation may result in very low attrition tomorrow, and that he therefore should continue as on the first day. Unfortunately for that commander, things won't "average out." He needs to change tactics.

So far the equations have represented input uncertainty. However, suppose that we do not know the correct equations of combat—except that, for some reason, we are convinced that the correct equations are Lanchesterian: either what aficionados call "Lanchester linear," "Lanchester square," or something in between. We could then reflect this uncertainty by rewriting the equation as

$$\frac{d\tilde{R}}{dt} = -\tilde{K}_b \tilde{B}^e(t)\tilde{R}^f(t) \quad \frac{d\tilde{B}}{dt} = -\tilde{K}_r \tilde{B}^g(t)\tilde{R}^h(t) \ .$$

Now, by treating the exponents e, f, g, and h as uncertain parameters, we can change the very structure of the model. Thus, by varying parameter values, we can explore both input and structural uncertainties in the model.

There are limits to what can be accomplished. Suppose that the correct equations of combat are indeed Lanchesterian but that the K-factors decay exponentially with time as combatants tire, lose efficiency, or husband ammunition. The consequences of different exponential decay times cannot even be explored if the phenomenon goes unrecognized. This is not an idle comment; we often do not know the underlying form of the system model: many aspects of phenomena are recognized, but not others. And they may not be observed except in unique circumstances. Despite this caveat, we can do a great deal with exploratory analysis to understand the consequences of uncertainties that can be parameterized.

Chapter Ten

USING EXPLORATORY MODELING

Daniel B. Fox

This chapter examines a way to use combat modeling that both capitalizes on the strengths of combat models and helps analysts and decisionmakers gain new insights into complex problems.

The chapter has three sections. The first of these describes the evolving defense environment, to show the need for tools that allow analysis of situations dominated by uncertainty. It also briefly discusses combat models in general and the Joint Integrated Contingency Model (JICM) in particular, covering some features of the JICM that make it especially suitable for exploratory modeling. Finally, the section compares conventional sensitivity analysis to exploratory modeling.

The second section describes how exploratory modeling is done, discussing its experimental design and its measures of outcome. It also presents a comprehensive example of applied exploratory modeling, identifying the problem and illustrating some conclusions. The third section then briefly describes exploratory modeling's key advantages.

THE NEED FOR EXPLORATORY MODELING

This book underscores the fact that the national security environment has changed dramatically and continues to change. During the Cold War, the role of U.S. military forces was to prepare for a major war in Central Europe. Other requirements—preparing to fight in Korea or in smaller-scale conflicts—were considered "lesser included cases" of the requirement for Central Europe. With the fall of the Berlin wall, the role of the U.S. military is now to prepare for a variety of contingencies, including terrorist threats. Although the nature of the future defense environment is unclear, long-term deployments to rescue "failed states" such as Afghanistan seem more likely, while having to commit forces to a major theater war (MTW) appears less so. These smaller but more likely operations are difficult to analyze. They are not amenable to many analytic tools and are dominated by uncertainties, ranging from the nature of the conflict, to the location, to the possible reactions to U.S. actions taken in response to evolving conditions.

At the same time, despite the stunning immediate aftermath of the September 11 attacks, many of these contingencies or deployments, in and of themselves, will invoke only limited national interests. The use of overwhelming force is one way to limit casualties, but sending large force deployments to problem areas stresses the rotation base of the services. The culmination of these stresses is a push for new ways of pursuing national security interests, which, in turn, creates a demand to analyze new alternatives.

Exploratory analysis is a tool to aid decisionmaking in such uncertain environments. It applies combat modeling to analytic problems in ways that have not been widely used. In particular, it uses the enormous computation capabilities of modern computers to intensively explore alternative outcomes by systematically varying assumptions. Paul Davis's Chapter Nine describes the technique in great detail. This chapter provides concrete illustrations of how this powerful technique can be used.

Given both the limitations and utility of models, it remains true that model-aided analysis says more about the analyst than about the model. It is the analyst who must judge how to represent the myriad details of the situation under study. In a nutshell, all combat models are wrong. But some, in conjunction with intelligent analysts, can be useful.

The Joint Integrated Contingency Model

The RAND-developed JICM is one such useful model. It employs modular functional submodels (some of which are listed in Figure 10.1) to manipulate the objects represented within the overall model.

Figure 10.1—JICM Functional Submodels

In JICM, one functional submodel is the simulation's strategic mobility module, which allows the analyst to set up simulation experiments that explicitly include enemy actions intended to degrade U.S. mobility. The degradation in mobility causes adjustments to the arrival of U.S. forces, which, in turn, can affect downstream theater-level outcomes.

Most theater combat models use a scenario input file that is a linear presentation of the events to be simulated.[1] By contrast, JICM uses analytic war plans that explicitly implement the major operational-level decisions of the campaign and allow the campaign to develop along alternative paths in accordance with how the simulated situation evolves. Within JICM, the war plans can query the state of the simulation and then alter actions taken by entities in the simulation

---

[1]The linear presentation generally describes the major operational events in terms of the fixed time when they are to occur in the simulation. Such events include the arrival of forces, and the timing of offensive and counteroffensive actions.

based on the results of the queries. Three examples of this kind of query are

1. If ?control[KuwaitCity]==Iraq then "do not use POMCUS"

2. If ?location[1-CAV/1-BDE]==KuwaitCity then "implement delay"

3. If (?tooth[EUSA] > 600 && ?tail[EUSA] > 800) then "begin CO"

Query 1 checks to see if Iraq has gained control of (the JICM place) Kuwait City. If so, the analytic war plans select a set of orders that does not involve attempting to use prepositioned combat equipment there. Query 2 verifies that a specific early-arriving force has arrived at Kuwait City. If so, the analytic war plans select a set of orders that implements actions to delay the advance of enemy forces in order to provide time for additional forces to arrive. Query 3 verifies that sufficient combat force and support ("tooth" and "tail") have arrived to begin the counteroffensive.

In JICM analysis, a single analytic war plan can include enough logic to react to the major operational turning points of a conflict. There is no need to create individual linear-order sequences for each alternative case to be examined.

Sensitivity Analysis and Exploratory Modeling

As is true for most models, the use of combat models typically involves some form of sensitivity analysis. In basic form, sensitivity analysis consists of three steps:

1. Establish a base case and obtain results

2. Define an alternative case by changing one or more input variables and obtain new outcomes

3. Compare the base case and alternative case, repeating steps 2 and 3 as required.

In contrast to sensitivity analysis, exploratory analysis is a more intensive process in which a range of values for a set of input variables is defined. Exploratory analysis then executes the simulation for every combination of values for all variables. Full enumeration of all possible cases can quickly mushroom to a very large number of runs.

Varying the numbers of variables and the number of values assigned to those variables produces numbers of runs for conventional sensitivity and exploratory modeling as follows:

- Conventional sensitivity: To explore sensitivity to n variables with m values each, the experiment size is 2 raised to the nth power and the number of runs is thus, e.g.,

  32 with n = 5, m = 2

  1,024 with n = 10, m = 3

  1,048,576 with n = 20, m = 4.

- Exploratory modeling: To explore an experiment with n variables with m values each, the experiment size is m^n and the number of runs is thus, e.g.,

  32 with n = 5, m = 2

  59,049 with n = 10, m = 3

  greater than a trillion with n = 20, m = 4.

Figure 10.2 shows the hours or computers needed for simulation runs. The top half of the figure shows how many hours it takes for a specified number of runs (10 to 1,000,000, across the columns) as a function of the time for each simulation run (3 to 3,000 minutes, down the rows). Networks of computers are now routinely available, so the lower half of the figure converts to an alternative metric: how many computers are needed to execute the specified number of simulations within a reasonable time limit (one week).

The number of exploratory modeling cases expands quickly as the number of variables and values rises. Such large numbers can easily tax the computation limits of even large networks of modern computers. Although most simulations are "fast" when running a single case, execution time becomes critical when exploratory modeling requires that thousands of cases be run. Thus, the art of exploratory modeling is being able to limit the analysis to the most important cases. To do so, some conventional sensitivity analysis might be used prior to the exploratory modeling in order to identify important variables in the decision space.

Number of simulation runs

Minutes per simulation run	10	100	1,000	10,000	100,000	1,000,000	
3	1	5	50	500	5,000	50,000	
30	5	50	500	5,000	50,000	500,000	Total hours
300	50	500	5,000	50,000	500,000	5,000,000	
3,000	500	5,000	50,000	500,000	5,000,000	50,000,000	
	10	100	1,000	10,000	100,000	1,000,000	
3	1	1	1	3	30	300	Number of computers to complete simulations in 1 week (168 hours)
30	1	1	3	30	300	3,000	
300	1	3	30	300	3,000	30,000	
3,000	3	30	300	3,000	30,000	300,000	

Figure 10.2—Hours or Computers Required for Simulation Runs

DOING EXPLORATORY MODELING

Apart from its dependence on the validity of the combat model itself, exploratory modeling rests on two fundamentals: experimental variables and measures of outcome.

The selection of *experimental variables* and their assigned values constitutes the experimental design for an exploratory analysis. There is no general rule for identifying the best variables or values in a given circumstance; selection depends on the nature of the problem under study and the operational experience of the analyst(s) conducting the experiment. But the ranges selected for the variables should make an analytic difference. Experimental variables come in two types. Some are quantities that can, in some sense, be controlled in the real world (e.g., the quantity of force to be applied in numbers of divisions or squadrons), whereas others represent "risks," or some uncertainty that might require some form of hedge, or "insurance."

Measures of outcome are the experimental results used to determine the relative goodness of cases. The chosen measures of outcome should be operationally meaningful to decisionmakers and, at the same time, highlight differences between the cases in the experimental design.

An example can clarify the process for and problems in conducting an exploratory analysis. Consider an exploratory analysis examining a Southwest Asia (SWA) scenario that starts with an Iraqi attack through Kuwait into Saudi Arabia. Enemy activities and allied decisions have the potential to restrict U.S. access to the theater early in the conflict. Given these potential restrictions, three different U.S. force enhancements are to be assessed.

Four of the variables in this example represent risks. Two of the four represent enemy-controlled factors (mines and chemicals), the third represents a factor controlled by U.S. allies (political access limits), and the fourth represents a risk neither fully under enemy control nor subject to U.S. choice (actionable warning time). Three additional variables represent potential U.S. force alternatives—Naval brilliant antitank (NBAT), Army brilliant antitank (ABAT), and Sea Cavalry (SCAV). These seven variables are summarized as follows:

- Mines = number of days Strait of Hormuz closed.

- Chemicals = days of effect on tactical air sortie rates and airport of debarkation (APOD) and seaport of debarkation (SPOD) unload times.

- Warning = days between day on which U.S. forces begin mobilizing (C-Day) and day on which war begins (D-Day).

- Access = base, some, less, worst, where base = NATO and Gulf Cooperation Council (GCC) access on C-Day; some = Kuwait, United Arab Emirates (UAE), and NATO on C-Day, and other GCC on D-Day; less = Kuwait, UAE, and United Kingdom (UK) on C-Day, all others except Saudi Arabia on D-Day, and Saudi Arabia on D+2; and worst = Kuwait and UK on C-Day, all others except Saudi Arabia on D-Day, and Saudi Arabia on D+4.

- NBAT = Naval-based ATACMs, 300 ship-based brilliant antitank (BAT) missiles that can be fired beginning on D-Day.

- ABAT = Ground-based BAT missile launchers stationed in the theater and 500 missiles available as soon as airlift can move missiles to the theater.

- SCAV = Sea Cavalry, ship-based attack helicopter concept providing for 1,000 sorties over the period D+0 to D+9.

A comprehensive assessment of this scenario might include an examination of various measures of outcome, but for illustrative purposes, this exposition considers only one measure—the maximum depth of penetration of enemy forces into friendly territory. Coding the measure of outcome permits rapid examination of multiple alternative cases by showing the results of several simulation experiments on a single diagram. The coding for the sample assessments is illustrated in Figure 10.3, where the darker the shading, the deeper the enemy penetration.

The encoded outcomes for 12 simulation runs are shown in Figure 10.4. Along the x axis are four different values for the access variable; along the y axis are three different values for the warning time. The values for all remaining variables are shown to the right of the illustration.

Examining Figure 10.4 in more detail highlights the fact that when there is zero warning, enemy forces penetrate deeply even under the least restrictive access constraints (no enemy mines or chemical weapons used).

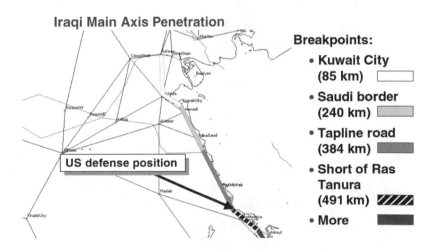

Figure 10.3—Coded Measure of Outcome

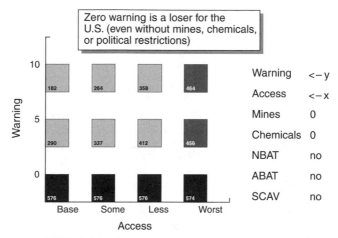

NOTE: The actual number of kilometers of
enemy penetration is shown in the lower left-
hand corner of each display box.

Figure 10.4—Exploring Access and Warning

Figure 10.5 introduces an additional display axis to augment the information of Figure 10.4. The shaded boxes appear to be stacked in sets of three; each set represents three different values of the mine variable along a z axis (drawn to allow the simultaneous presentation of more cases). Examining Figure 10.5 reinforces the previous observation that when there is zero warning time, the enemy forces are able to penetrate deeply.

Figure 10.6 replaces the mine variable on the z axis of Figure 10.5 with the chemical variable. Comparing the two suggests that chemicals may have a somewhat greater impact than mines do. Figure 10.7 directly compares mines and chemicals by moving the access variable to the z axis and putting chemicals and mines on the x and y axes. Here, warning time is five days.

Figure 10.7 makes it clearer that the outcomes tend to worsen faster as the chemical effects increase (as one moves to the right on the x axis) than they do as the mine effects increase (moving up on the y axis). That is, step increases in the chemical threat allow for greater

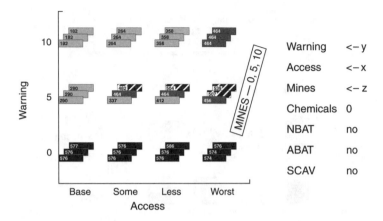

Figure 10.5—Exploring Access, Warning, and Mines

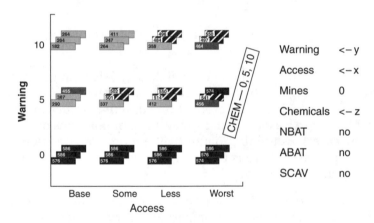

Figure 10.6—Exploring Access, Warning, and Chemicals

increases in enemy penetration than do step increases in the mine threat.

Figure 10.8 compares all three force alternatives—ABAT, SCAV, and NBAT. An examination of enemy penetration (measured in kilometers and displayed in the lower left corner of each display box) suggests that the ABAT or SCAV option restricts enemy penetration far

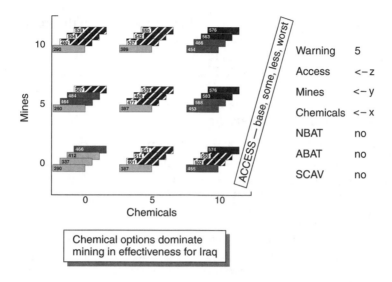

Figure 10.7—Exploring Chemicals, Mines, and Access

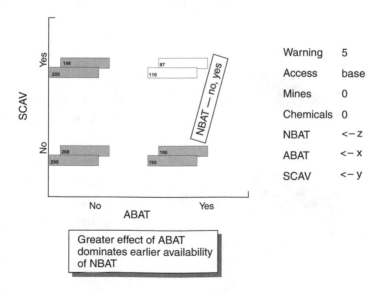

Figure 10.8—Exploring ABAT, SCAV, and NBAT

more than the NBAT option does. If the display box in the front lower left of Figure 10.8 is taken as a base case, NBAT alone reduces enemy penetration by roughly 20 km (comparing the front and rear display boxes in the lower left). ABAT alone saves 90 km (comparing the front boxes in the lower left and lower right); SCAV alone saves 80 km (comparing the front boxes in the lower left and upper left). The greater effect of ABAT offsets the earlier availability of NBAT.

Figure 10.9 illustrates how robust the ABAT and SCAV options are under the stress of enemy chemical actions. Comparing the two (i.e., comparing the three values in the display boxes in the upper left with the three values in the boxes in the lower right) suggests that they reduce enemy penetration to a similar degree.

Looking at the three values stacked along the z axis of Figure 10.9, one can see that the advantages of ABAT and SCAV are reduced when the enemy uses chemical weapons. At worst, the enemy penetrates 550 km. If both ABAT and SCAV are available (display boxes in upper right), the outcomes are substantially improved over those from either option on its own, suggesting that the two reinforce each other.

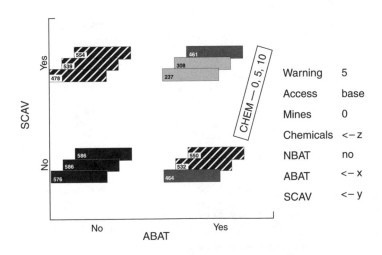

Figure 10.9—Exploring ABAT and SCAV When Chemicals Are Used

Figures 10.4 through 10.9 represent just a few of the exploratory analysis displays possible with this experimental design. In practice, any of the experimental variables can be displayed on the x, y, and z axes, and any variable not on an axis can be set to any desired value. In addition, any of the collected measures of outcome can be displayed. The figures provided here represent just one possible line of exploration through the experimental space.

THE VALUE OF EXPLORATORY MODELING

Exploratory modeling not only examines risks and force alternatives; it also can be used to test the effect of alternative theater concepts of operation (CONOPs) and alternative investments in mobility (e.g., prepositioning ashore, prepositioning afloat, and greater numbers of mobility ships or aircraft). Additionally, it permits a more extensive assessment of the U.S. force structure's robustness by making it possible to examine the many factors that might make the U.S. defense case more difficult (e.g., insufficient warning, a lack of allied contributions, unanticipated enemy strength, shortfalls in critical ammunition).

Some of JICM's features make it particularly suitable for exploratory modeling at the theater level. They include the breadth of scenarios that can be represented, the ability to address strategic mobility, and a flexible war plan system that permits many variations of the basic cases to be created.

Another advantage of exploratory modeling is that it is not limited to theater combat problems. It may be applied wherever a suitable simulation model can run all the cases in an experimental design in a reasonable time. Thus, for example, it is suitable for analyzing smaller-scale contingencies in which combat outcomes may not be the defining feature, environmental degradation, or traffic management.

Regardless of the problem to be studied, a key advantage of exploratory analysis is the ability to model both uncertainty—by using variables to represent things not under decisionmakers' control—and alternative choices. In using a model, the analyst is forced to organize all thoughts about the problem. An exploratory analysis can provide a rich illustration of the effects alternatives will have under varying circumstances, thus permitting a full appreciation of the choices.

Chapter Eleven

ASSESSING MILITARY INFORMATION SYSTEMS

Stuart H. Starr

The assessment of military information systems is an art form that has evolved substantially since the mid-1970s.[1] Until then, national security assessments generally neglected information system issues. Such systems were either assumed to be "perfect" (i.e., providing perfect information with no time delays), capable of no more than second- or third-order effects, or totally irrelevant. When they were considered, they were often treated as a "patch"—something introduced into force-on-force calculations to reflect imperfect information systems.

This chapter begins with a historical perspective on how military information systems assessment evolved. It describes the change that took place 25 years ago, which entailed a basic reengineering of the assessment process, one that involved integrating leadership, institutions, people, processes, resources, tools, research and development (R&D), and products. An initial period of innovative assessments of military information systems was followed by a hiatus in the 1980s as Department of Defense (DoD) leadership lost interest in analyzing information systems. "Paralysis through analysis" was an oft-heard criticism in the Pentagon. Budgets for military systems, including military information systems, were at historically high levels, and the Pentagon's emphasis was on acquiring military information systems, not assessing them.

---

[1]Stated more precisely, this would read "information systems in support of military operations, including systems owned and operated by nonmilitary organizations." These include command and control centers, communications, sensors, and ancillary systems (e.g., navigation).

The chapter then addresses how this attitude shifted in the early 1990s, thanks in large part to profound changes in the international scene. The Soviet Union dissolved and the Persian Gulf War provided an insight into how innovative military information systems could support contemporary warfare. Thus emerged new challenges in assessing military information systems, and hence a new information assessment process. The chapter then moves to the key principles and insights related to the assessment of information systems in the context of conventional conflict. These insights are encapsulated in the 1999 *NATO Code of Best Practice for Command and Control Assessment*,[2] a product of six years of deliberations by nine NATO nations.

The chapter concludes by summarizing the advances made in the art of assessing military information systems since the mid-1970s, and then turning to the challenges that remain, such as the development and implementation of novel assessment tools and the treatment of emerging "new world disorder" missions (i.e., coercive operations using a mix of diplomatic, informational, economic, and military resources to convince an adversary to withdraw military forces from a neighbor), and peacekeeping, homeland defense, counterterrorism, counter-weapons of mass destruction (WMD), and information operations.

HISTORICAL PERSPECTIVE

Figure 11.1 summarizes the factors that fundamentally changed how military information systems were assessed from 1975 to 1985. Key civilian and military leaders in the defense community—Robert Hermann (then assistant secretary of the Air Force for Research, Development, and Acquisition), Harry Van Trees (then principal deputy assistant secretary of defense for Command, Control, Communications, and Intelligence), Charles Zraket (then executive vice president, MITRE), and MG Jasper Welch (then director, Air Force Studies and Analyses), and others—launched a search for the "Holy Grail,"

---

[2]*NATO Code of Best Practice (COBP) on the Assessment of C2,* Research & Technology Organisation (RTO) Technical Report 9, AC/323(SAS)TP/4, Communication Group, Inc., Hull, Quebec, March 1999. (Text also available at http://www.dodccrp.org/nato_supp/nato.htm.)

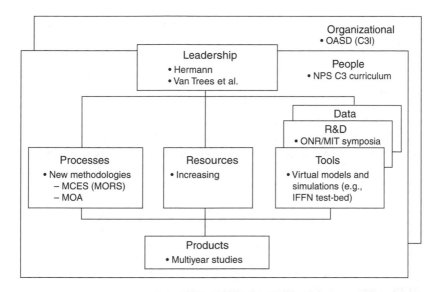

Figure 11.1—A Business Process Reengineering Perspective of Information Assessment (1975–1985)

i.e., for the ability to assess the impact of C2 systems on force effectiveness.[3] They acted out of intellectual curiosity, but also because of their emerging awareness of the importance of military information systems in modern warfare and their need to justify budgets for military information systems to a skeptical Congress.

This initiative was helped by the creation of the office of the assistant secretary of defense (OASD) for command, control, communications, and intelligence (C3I), an action that brought together the disparate DoD organizations responsible for C2, communications, intelligence, and defense support systems (e.g., electronic warfare, navigation). The contemporary establishment of a C3 curriculum by the Naval

---

[3]Major General Jasper A. Welch, Jr., "Command and Control Simulation—A Common Thread," keynote address, *AGARD Conference Proceedings*, No. 268 ("Modeling and Simulation of Avionics Systems and Command, Control and Communications Systems"), Paris, France, October 15–19, 1979; and OSD, with the cooperation of MITRE Corporation, C3 Division, *Proceedings for Quantitative Assessment of Utility of Command and Control Systems*, National Defense University, Washington, DC, January 1980.

Postgraduate School (NPS) helped create the human capital needed to assess military information systems. Finally, the Office of Naval Research (ONR) established a multiyear program with the Massachusetts Institute of Technology (MIT) to pursue R&D on information system assessment; the principals were innovators in the field of optimal control systems. Although the optimal control paradigm proved to have only limited applicability to the major issues associated with information systems, it helped address a key subset of them (e.g., the multisensor, multitarget fusion problem). The program built a vibrant community of interest that acquired a shared understanding of the problem.

Several new methods for assessing information systems consequently emerged. Workshops sponsored by the Military Operations Research Society (MORS) spawned the Modular Command and Control Evaluation Structure (MCES), a framework for defining and evaluating measures of merit for assessing information systems.[4] This framework was subsequently adapted and extended by the NATO COBP (see below).

In the mid-1980s, the "mission oriented approach" (MOA) to C2 assessment was developed and applied; its key phases are summarized in Figure 11.2.[5] The approach addresses four questions:[6]

1. What are you trying to achieve operationally?

2. How are you trying to achieve the operational mission?

3. What technical capability is needed to support the operational mission?

4. How is the technical job to be accomplished?

---

[4]Ricki Sweet et al., *The Modular Command and Control Evaluation Structure (MCES): Applications of and Expansion to C3 Architectural Evaluation*, Naval Postgraduate School, September 1986; and Ricki Sweet, Morton Metersky, and Michael Sovereign, *Command and Control Evaluation Workshop* (revised June 1986), MORS C2 MOE Workshop, Naval Postgraduate School, January 1985.

[5]David T. Signori, Jr., and Stuart H. Starr, "The Mission Oriented Approach to NATO C2 Planning," *SIGNAL*, September 1987, pp. 119–127.

[6]The questions are posed from the perspective of the friendly coalition, which subsumes operational users, military information systems architects and developers, and the science and technology community.

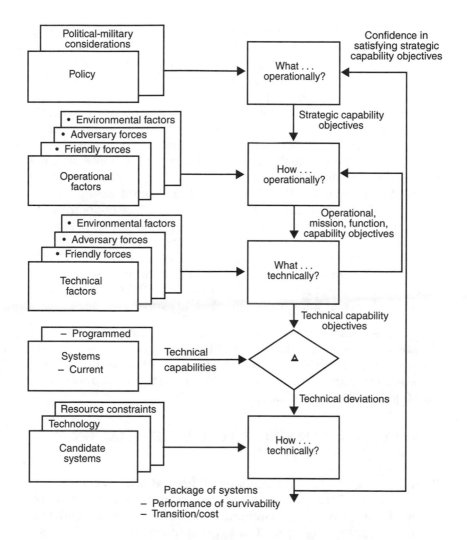

Figure 11.2—Phases of the Mission Oriented Approach

The MOA drove home the importance of assessing military information systems within the context of the missions they support. The process is implemented through top-down decomposition linking missions, functions, tasks, and systems. The "roll-up" process—in which analysts assess how candidate packages of military information systems satisfy mission objectives—remains a challenge.

One challenge of the roll-up process is understanding the performance of distributed teams of operational users under stress. To address this issue, manned simulator test-beds were developed to represent the specific weapons systems and military information systems supporting selected missions. An early example, the Theater Air Command and Control Simulation Facility (TACCSF) (originally, the Identification Friend Foe or Neutral [IFFN] test-bed), brought together teams of operators manning simulated weapons systems (e.g., airborne interceptors, high- to medium-range air defense systems) and associated military information systems (e.g., Airborne Warning and Control System [AWACS], associated ground-based C2 systems).[7] Such test-beds can flexibly assess a full range of doctrine, organization, training, materiel, leadership and education, personnel, and facilities issues associated with systems-of-systems. (This combination of factors is characterized in *Joint Vision 2020*[8] by the infelicitous initialism "DOTML-PF".) Recent advances in computer science—e.g., the High Level Architecture (HLA)—have greatly helped such virtual simulations emerge and evolve.

New studies drew on these methods and tools to provide logical, systematic links between packages of military information systems and overall mission effectiveness. An early example of these products was developed as part of the NATO C3 Pilot Program in support of the Tri-Major NATO Commanders C3 Master Plan.[9]

CONTEXT FOR ASSESSING MILITARY INFORMATION SYSTEMS IN THE 21st CENTURY

Table 11.1 highlights some dramatic shifts that have occurred since the end of the Cold War. As is now commonplace to observe, the Soviet Union provided a sustained focus for intelligence gatherers and force planners during the Cold War. A few scenarios and types of

---

[7]J. E. Freeman and S. H. Starr, "Use of Simulation in the Evaluation of the IFFN Process," Paper 25, *AGARD Conference Proceedings*, No. 268 ("Modeling and Simulation of Avionics Systems and C3 Systems"), Paris, France, October 15–19, 1979.

[8]*Joint Vision 2020*, Chairman of the Joint Chiefs of Staff, Director for Strategic Plans and Policy, J5, Strategy Division, U.S. Government Printing Office, Washington, DC, June 2000.

[9]K. T. Hoegberg, "Toward a NATO C3 Master Plan," *SIGNAL*, October 1985.

Table 11.1

A New DoD Context

Area	Old	New
Threat	Relatively well-understood	New, uncertain
Missions	Established scenarios and operations	Broader range
Focus	DoD, alliance	National, coalition
Capability	Evolutionary	Revolutionary
Force	Overwhelming	Information/effects-based
Advantage	System-on-system	System-of-systems
Requirements	Relatively well-defined	Exploration/learning

operations sufficed for assessment and planning. Today's broader range of uncertain threats has made it difficult to anticipate issues and focus intelligence resources appropriately. The United States faces a variety of "new world disorder" missions, as well as the more conventional military missions (e.g., smaller-scale contingencies and major theater wars [MTWs]).

Alliances aside, DoD used to concern itself mainly with operations that involved only the U.S. military. Today's operations usually involve many other participants, such as ad hoc coalitions, various national government organizations, international organizations, and nongovernmental organizations (NGOs). Hitherto, warfighting capability evolved incrementally with the addition of each new weapons system. Now, information technology and precision weaponry may well change the nature of warfare in revolutionary ways.

Previously, success was thought to be determined by who could bring to bear overwhelming force. Today, the U.S. goal is to gather and exploit information about its adversaries so as to be able to apply the minimum force needed to achieve a specific effect, consistent with national policy. This force often transcends purely military action to include diplomatic, informational, and economic means. Advantage used to be measured in platform-centric terms—who had the best tank, ship, or plane. Today, networking sensors, C2, and weapons in a system-of-systems promise significant advantage through increased agility and more-discriminate application of force.

Finally, the stable, evolutionary environment of relatively well understood requirements has yielded to a period of experimentation and learning directed at understanding how to exploit new technologies and new concepts for competitive advantage. All told, such shifts mean a fundamentally different national security context for today's analysts, especially for those who assess information systems that play a newly critical role in force transformation. Table 11.2 highlights some of the key changes.

Analysts once could focus on ways to counter a specific threat; today, they must address capabilities that can be used in an agile manner to deal with a range of threats.[10] The stability of the threat and the evolutionary nature of military capability once permitted analysts to focus on refining established operational concepts and capabilities; today, they must explore completely new warfighting concepts, such as distributed C2 for a nonlinear battlespace. Cold War analysts could focus on the benefits of adding a new weapons system to the force mix; today's analysts must understand the fundamentals associated with networking the force or sharing information through a common relevant operational picture.

In the recent past, assessments focused on force mix/structure issues. Now, they must address *all* the elements of doctrine, organization, training, materiel, leadership and education, personnel, and

Table 11.2

New DoD Assessment Challenges

Area	Old	New
Planning	Threat-based	Capability-based
Focus	Refine established notions	Explore new possibilities
Objective	Identify benefits of incremental, new capabilities	Understand fundamentals
Assessment	Force structure	DOTML-PF
Issues	Ad hoc collection	Hierarchy of related issues
Complexity	Tractable	Exploding

[10]U.S. Department of Defense, *2001 Quadrennial Defense Review Report*, September 30, 2001, p. iv.

facilities (DOTML-PF). Whereas analysts used to concentrate on ad hoc issues arising in the programming and budgeting processes, they now must make systematic multilevel assessments of a comprehensive issue set.

Moreover, military information systems are themselves growing more complex. Table 11.3 highlights how emerging systems-of-systems and other integration proposals (e.g., the global information grid [GIG]) add further complexity. The challenge of assessing such systems reminds one of John Von Neumann's maxim: "A system is complex when it is easier to build than to describe mathematically."[11]

ADDITIONAL COMPLICATING AND SUPPORTING FACTORS

In assessing tomorrow's military information systems, additional factors must be considered: those that complicate the task (particular initiatives, especially by the services, and system trends) and those that can assist the analyst (such as new tools and new kinds of workshops).

Table 11.3

Simple Versus Complex Systems

Attributes	Simple Systems	Complex Systems
Number of elements	Few	Many
Interactions among elements	Few	Many
Attributes of elements	Predetermined	Not predetermined
Organization of interaction among elements	Tight	Loose
Laws governing behavior	Well-defined	Probabilistic
System evolves over time	No	Yes
Subsystems pursue their own goals	No	Yes
System affected by behavioral influences	No	Yes

SOURCE: R. Flood and M. Jackson, *Creative Problem Solving,* John Wiley, New York, 1991.

---

[11]J. Von Neumann, *Theory of Self-Replacing Automata,* University of Illinois Press, Urbana, IL, 1996.

Each service has undertaken activities to transform how it will operate.[12] The U.S. Army is transforming itself via the Stryker Brigade
Combat Team (SBCT), the Future Combat System (FCS), and the
Objective Force Warrior (OFW) to enhance deployability, sustainability, lethality, and survivability. These initiatives aim to dominate
potential adversaries across the full conflict spectrum using information systems as the key force multiplier. The U.S. Air Force is creating
an Expeditionary Aerospace Force, with enhanced responsiveness
and global reach. These objectives are pursued through enhanced
reach-back capability (e.g., using substantial resources in sanctuary
to reduce the footprint in theater) and advanced collaboration tools
(e.g., implementing the "virtual building" paradigm[13]). The U.S.
Navy is pursuing the doctrine articulated in "Forward from the Sea,"
through the concept of network-centric warfare. Moving away from a
platform-centric approach calls for the co-evolution of all the
components of DOTML-PF (i.e., self-consistently modifying all
aspects of the service's doctrine, organization, training, materiel,
leadership and education, personnel, and facilities). A network-centric focus is promoted to enhance mission effectiveness through
shared awareness and self-synchronization of the force. And, finally,
the U.S. Marine Corps is using experimentation to refine "Marine
Corps Strategy 21" through the innovative use of information systems to support small unit operations and urban warfare.

Information systems are perceived to be the key enablers for all these
initiatives. From a joint perspective, the J-9 organization of Joint
Forces Command (JFCOM) has an ambitious agenda to evaluate new
joint concepts enabled by the revolution in information systems.

There is also growing interest in developing a joint, integrated information system infrastructure to underpin the operations of all missions. One aspect would be to design interoperability and security
into the evolving systems-of-systems rather than treating them as
add-ons. These initiatives include the Office of the Secretary of
Defense (OSD)/Joint Staff efforts to have the GIG subsume the rele-

---

[12]*Service Visions*, available at http://www.dtic.mil/jv2020/jvsc.htm.

[13]Peter J. Spellman, Jane N. Mosier, Lucy M. Deus, and Jay A. Carlson, "Collaborative
Virtual Workspace," *Proceedings of the International ACM SIGGROUP Conference on
Supporting Group Work: The Integration Challenge*, 1997, pp. 197–203.

vant service initiatives: the Air Force's Joint Battlespace Infosphere[14] and the Army's Tactical Infosphere.[15] The systemwide initiatives would exploit the power of Web-based architectures, commercial standards and protocols, and emerging information system markup languages based on the extensible markup language (XML).

However, many of the existing assessment tools for information systems are a poor fit for this class of systems, so several joint efforts are under way to develop better ones: the Joint Warfare System (JWARS) for joint assessment and the Joint Simulation System (JSIMS) for joint training. Although these tools seek to reflect information systems explicitly (and, in several cases, to interface with operational information systems), they are still immature and largely restrict themselves to the issues associated with conventional warfare.

Over the last decade, workshops—notably those by MORS—have advanced the understanding of challenges and opportunities associated with assessing military information systems. Recent MORS workshops have sought to identify the shortfalls in the community's ability to assess the impact of information systems[16] and to formulate a plan of action to ameliorate these shortfalls.[17]

NATO CODE OF BEST PRACTICE

In 1995, NATO established Research Study Group 19 to develop a code of best practice (COBP) for assessing C2 in conventional conflict; that COBP was issued under the newly formed NATO Studies, Analysis, and Simulations (SAS) panel.[18] A follow-up effort is extending the COBP to assess C2 for operations other than war (OOTWs).

---

[14]USAF Scientific Advisory Board, *Report on Building the Joint Battlespace Infosphere, Vol. 1: Summary,* SAB-TR-99-02, December 17, 1999.

[15]Army Science Board 2000 Summer Study, "Technical and Tactical Opportunities for Revolutionary Advances in Rapidly Deployable Joint Ground Forces in the 2015–2025 Era," Panel on Information Dominance, July 17–27, 2000.

[16]Russell F. Richards, "MORS Workshop on Analyzing C4ISR in 2010," *PHALANX,* Vol. 32, No. 2, June 1999, p. 10.

[17]Cy Staniec, Stuart Starr, and Charles Taylor, "MORS Workshop on Advancing C4ISR Assessment," *PHALANX,* Vol.34, No. 1, March 2001, pp. 29–33.

[18]*NATO Code of Best Practice on the Assessment of C2.*

Figure 11.3 portrays major elements of an effective assessment process for information systems that was identified in the NATO COBP. It highlights the major steps of the assessment and the products that should be developed in it. Despite the fact that the following discussion reflects this framework, however, it should be noted that meaningful assessments of information systems rarely follow such a linear process. Recent experience suggests that the way to best fit the problem at hand is to tailor and implement a nonlinear process that iterates among these steps.

Before the assessment process begins, the first issue is who will participate. Such undertakings generally require interdisciplinary teams of individuals skilled in operations research, modeling and simulation, information systems, and operations. Extensions of the COBP to OOTW also highlights the need to include those skilled in social sciences (e.g., political science and demography). Once a team is established, the process proceeds as follows, in line with Figure 11.3:

Problem Formulation. Military information system issues are complex, poorly defined, and hard to formulate sharply—especially when used in OOTWs, where cultural and historical context must be understood. Such issues are also hard to decompose into pieces that can be analyzed individually and then brought together coherently. Worse, posing options in strictly materiel terms is rarely acceptable. Issues associated with military transformation must address all the dimensions of DOTML-PF.

Human Factors and Organizational Issues. Information systems generally support distributed teams of people operating under stress; changing these systems often leads to altered tactics, techniques, procedures, and DOTML-PF—all of which must be considered in the assessment. These issues are often difficult to assess and are the subject of ongoing research efforts.[19] Factors such as belief (e.g., morale, unit cohesion), cognitive processes (e.g., naturalistic decisionmaking), and performance modulators (e.g., fear, fatigue, and sleep deprivation) are especially challenging to address.

---

[19]William G. Kemple et al., "Experimental Evaluation of Alternative and Adaptive Architectures in Command and Control," *Third International Symposium on Command and Control Research and Technology*, National Defense University, Fort McNair, Washington, DC, June 17–20, 1997, pp. 313–321.

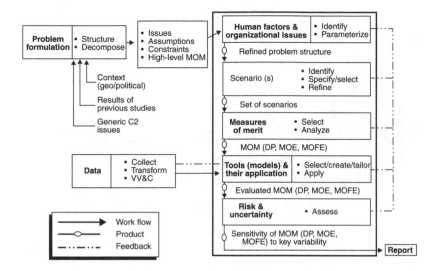

Figure 11.3—NATO COBP Assessment Methodology

Scenarios. The NATO COBP holds that military information systems can only be assessed relative to selected scenarios. Figure 11.4 identifies a scenario framework formulated in the NATO COBP; it is based on three major categories: external factors (e.g., political, military, and cultural situation), capabilities of actors (e.g., friendly forces, adversary forces, and noncombatants), and the environment (e.g., geography, terrain, and man-made structures). The challenge is to explore the scenario space rapidly and focus on its more "interesting" regions. Because military information systems are complex, looking at just one scenario is almost always a mistake. It is thus necessary to decompose the three major categories of the scenario framework, selecting a baseline scenario and interesting excursions.[20]

Measures of Merit. The NATO COBP states that no single measure exists by which the overall effectiveness or the performance of military information systems can be assessed. Drawing on prior MORS

---

[20]Stuart H. Starr, "Developing Scenarios to Support C3I Analyses," *Proceedings of the Cornwallis Group,* Pearson Peacekeeping Center, Nova Scotia, Canada, March 26–28, 1996.

External Factors	Political, military, and cultural situation	Mission objectives, mission constraints, rules of engagement	Mission tasks (e.g., military scope and intensity, joint/combined)
	National security interests		
Capabilities of Actors	• Organization, order of battle, C2, doctrine resources • Weapons, equipment • Logistics, skills, morale, etc.		
	Friendly forces	Adversary forces	Noncombatants
Environment	• Geography, region, terrain, accessibility, vegetation • Climate, weather • Civil infrastructure (e.g., transportation, telecommunications, energy generation/distribution)		

Figure 11.4—The Scenario Framework

workshops,[21] NATO recommended a multilevel hierarchy of measures of merit (MOMs), four levels of which are shown in Figure 11.5 and can be defined as follows:

- Measures of force effectiveness (MOFEs): how a force performs its mission (e.g., loss exchange ratios).

- Measures of C2 effectiveness (MOEs): impact of information systems within the operational context (e.g., the ability to generate a complete, accurate, timely common operating picture of the battlespace).

- Measures of C2 system performance (MOPs): performance of the internal system structure, characteristics, and behavior (e.g., timeliness, completeness, or accuracy).

- Dimensional parameters (DPs): properties or characteristics inherent in the information system itself (e.g., bandwidth).

Extending the NATO COBP to OOTW has demonstrated that the hierarchy of MOMs must be expanded to include measures of policy effectiveness. Since the military plays only a contributing role in such missions—ensuring that the environment is secure enough that

---

[21]Thomas J. Pawlowski III et al., *C3IEW Measures of Effectiveness Workshop*, Final Report, Military Operations Research Society (MORS), Fort Leavenworth, KS, October 20–23, 1993; and Sweet, Metersky, and Sovereign, "Command and Control Evaluation Workshop."

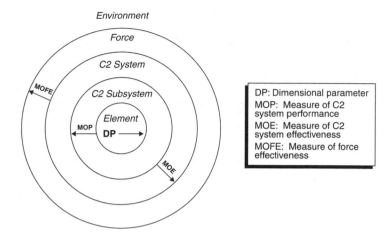

Figure 11.5—Relationships Among Classes of Measures of Merit

other organizations can function effectively—the contribution of international organizations and NGOs must be captured. Table 11.4 depicts representative MOMs for a hypothetical civil-military operations center (CMOC) that would provide the needed linkage between the military community and other organizations participating in the operation.

Historically, assessing what the MOMs at the top of the hierarchy implied for those measures at the bottom was a straightforward task. For instance, minimizing the leakage of incoming ballistic missiles creates a need for early warning to be extended and delays by military information systems to be minimized.[22] However, it is often more challenging to go "bottom-up" to estimate the effectiveness of mixes of weapons and information systems in the context of the operational scenario.

Data. At a MORS workshop in the late 1980s on simulation technology, Walt LaBerge, then principal deputy under secretary of defense (Research & Engineering), gave a presentation entitled "Without

---

[22]Signori and Starr, "The Mission Oriented Approach."

Table 11.4

Strawman MOMs for a Civil-Military Operations Center

Measures of policy effectiveness	Progress in transitioning from a failed state to a stable one (e.g., successful democratization and the ability to conduct a fair election; dealing with displaced persons and relocating displaced families)
Measures of force effectiveness	Ability of military to create and sustain a secure environment
Measures of C2 effectiveness	Quality of situational awareness and synchronization of effort
Measures of C2 performance	Ability to perform CMOC tasks and functions (e.g., time to complete a task)
Dimensional parameters	Communications (e.g., bandwidth and connectivity), automated data processing, support to personnel (e.g., quality and flexibility), collaboration tools (e.g., scalability, latency, and security)

Data We Are Nothing."[23] Unfortunately, the military's information system assessment community remains "data poor," despite repeated recommendations to establish a communitywide program to collect, transform, and verify, validate, and certify (VV&C) needed data. The problem is worse for OOTWs, because key information is controlled by NGOs (or even private corporations, such as insurance companies). Administratively, there is a need for a data dictionary/ glossary at the outset of an assessment and a strategy for enhanced data management.

Tools and Their Application. Table 11.5 depicts a spectrum of assessment techniques. It discriminates among the various techniques by characterizing how they account for the systems, people, and operations/missions of interest. For example, in virtual modeling and simulation (M&S), real people are employed, interacting with simulated systems, in the context of a simulated operation. Conversely, in live M&S, real people are employed, interacting with real systems, in the context of a simulated operation. The COBP concluded that no

---

[23]*Proceedings of SIMTECH 97*, 1987–1988 (available through MORS office, Alexandria, VA).

Table 11.5

Spectrum of Assessment Techniques

	Assessment Techniques				
Key Factors	Analysis	Constructive M&S	Virtual M&S	Live M&S	Actual Operations
Typical application	Closed form, statistical	Force-on-force models; communications system models	Test-beds with humans in the loop	Command post exercises; field training exercises	After-action reports; lessons learned
Treatment of systems	Analytic	Simulated	Simulated	Real	Real
Treatment of people	Assumed or simulated	Assumed or simulated	Real	Real	Real
Treatment of operations/missions	Simulated	Simulated	Simulated	Real or simulated	Real
Resources	Relatively modest	Moderate to high	High to very high	Very high	Extremely high
Lead time to create	Weeks to months	Months to years	Years	Years	N/A
Lead time to use	Weeks to months	Weeks to months	Weeks to months	Weeks to months	N/A
Credibility	Fair to moderate	Moderate	Moderate to high	High	Very high

NOTE: M&S = modeling and simulation; N/A = not applicable.

single assessment technique would suffice for many issues of interest. A proper strategy must select and orchestrate a mix of techniques consistent with issues at hand and real-world constraints (e.g., resources, lead time). As concepts such as "information superiority" and "decision dominance" have gained interest, so has the need for tools to represent both friendly and adversary information processes. The need for discipline in applying these tools suggests that formal experimental design matrices should be employed to govern their application and support the generation of appropriate response surfaces.[24] Fast-running tools can filter down to interesting segments of solution space, at which point fine-grained tools (e.g., virtual models and simulations) can provide more-focused, in-depth assessments.

Tools that have been formally verified, validated, and accredited are, of course, preferred, but few tools have undergone such stringent quality control processes. Confidence in results thus arises when independent assessments using varying tools nonetheless reach consistent findings. As an example, to provide an initial "cut" at a complex problem, analysts are beginning to develop and employ system dynamics models. These models (e.g., the C4ISR Analytic Performance Evaluation [CAPE] family of models[25]) evolve from influence diagrams that characterize factors such as model variables, inputs, outputs, and system parameters. They capture information system performance by explicitly representing sensors of interest, aggregate aspects of C3 (e.g., explicit constraints on communications capacity; time delays experienced by C2 nodes), and the phases of the intelligence cycle. CAPE was employed in OSD's C4ISR Mission Assessment to characterize the information systems that supported the engagement of time-critical targets.[26] Figure 11.6 depicts a representative output from CAPE characterizing the sensor-to-shooter string by estimating the probability of placing time-critical targets at risk as a function of target type (i.e., range and frequency of relocation).

Agent-based modeling represents another way to explore a solution space rapidly. It adopts a bottom-up approach to operations modeling by characterizing individual behavior (e.g., response to live or injured friendly or adversary entities; reaction to friendly or adversary objectives) and deriving emergent behavior from the resulting

---

[24]Starr, "Developing Scenarios."

[25]Jeremy S. Belldina, Henry A. Neimeier, Karen W. Pullen, and Richard C. Tepel, *An Application of the Dynamic C4ISR Analytic Performance Evaluation (CAPE) Model*, MITRE Technical Report 98W4, The MITRE Corporation, McLean, VA, December 1997.

[26]Russell F. Richards, Henry A. Neimeier, W. L. Hamm, and D. L. Alexander, "Analytical Modeling in Support of C4ISR Mission Assessment (CMA)," *Third International Symposium on Command and Control Research and Technology*, National Defense University, Fort McNair, Washington, DC, June 17–20, 1997, pp. 626–639.

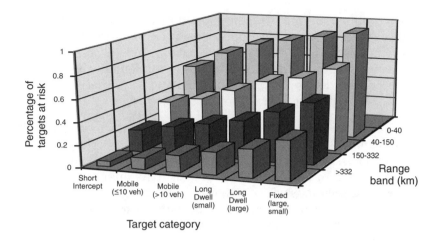

Figure 11.6—Representative CAPE Output: Targets at Risk

interactions.[27] Mana, a recent agent-based model developed by New Zealand's Defence Operational Support Establishment to prepare forces for peacekeeping operations in East Timor, has been employed to assess the risk to friendly personnel associated with alternative courses of action in OOTWs.[28]

Once the interesting parts of scenario space are identified, more-detailed simulations can explore them in greater depth. For example, to support the assessment of the time-critical target problem, the U.S. Defense Modeling and Simulation Organization (DMSO) developed Pegasus, a combination of three constructive simulations: the Extended Air Defense Simulation, Eagle, and the Navy Simulation System. JFCOM plans to use Pegasus and CAPE in its Model-Experiment-Model paradigm.

---

[27]Andrew Iiachinski, "Irreducible Semi-Autonomous Adaptive Combat (ISAAC): An Artificial-Life Approach to Land Combat," *MOR Journal*, Vol. 5, No. 3, 2000, pp. 29–46.

[28]Edward Brady and Stuart Starr, "Assessing C3I in Support of Dismounted Operations in Complex Terrain," *Proceedings of C2R&T Symposium*, NPS, Monterey, CA, June 11–13, 2002.

Other combinations are being developed that rely on virtual simulations to capture operators' response to a variety of stimuli. The Army is developing the Joint Virtual Battlespace (JVB) to assess and compare concepts proposed by contractors to implement the FCS, and the Air Force is developing the Joint Synthetic Battlespace to provide a context for acquiring a system-of-systems.

Risk and Uncertainty Assessment. The COBP notes that sensitivity analysis and risk assessment in C2 analyses often lack thoroughness because the issues are complex and the time and resources too limited. The need for and the results of sensitivity analyses should be stressed in discussions with decisionmakers. Analysts should at least test the robustness of the results against small excursions in the selected regions of scenario space. Ultimately, analysts must illuminate uncertainty, not suppress it.

Decisionmakers are also increasingly interested in getting risk-based instead of cost-benefit assessments. For example, the legislation mandating the 2001 Quadrennial Defense Review (QDR) specifically cast several questions in risk-based terms. This was echoed in the 2001 QDR itself, which introduced a new, broad approach to risk management.[29] The analysis community should also draw on the experience that financial planners and insurance actuaries have amassed in risk assessment.

This is the end of the process depicted in Figure 11.3. But, as discussed earlier, recent experience suggests that such a linear process rarely fits the needs of the situation. The last point is thus that an *iterative approach* is needed, since one pass through the assessment process is unlikely to generate meaningful answers. The first cut should be broad and shallow to identify key issues and relevant segments of scenario space, and subsequent iterations should then go narrower and deeper (drawing on suitable tools) to gain insight into key questions. Throughout this process, peer review is essential to provide adequate quality control.

---

[29] *2001 Quadrennial Defense Review Report,* Chapter VII, "Managing Risks," September 30, 2001.

ADVANCES OVER THE PAST 25 YEARS

Advances in the ability to assess military information systems are apparent in four areas. First and foremost, decisionmakers are now keenly aware that meaningful national security assessments require explicit consideration of military information systems. This awareness is apparent in recent products from the chairman of the Joint Chiefs of Staff, in which first "information superiority" and then "decision superiority" were placed at the foundation of DoD's strategic vision.[30] The easy, unknowing assumption that "information systems are perfect" is no longer acceptable.

Second, the processes for assessing military information systems have improved—a result of workshops (particularly those of MORS), individual studies (e.g., OSD's C4ISR Mission Assessment and Information Superiority Investment Strategy), and panels. Recent NATO panels have synthesized earlier efforts, promulgated COBPs, and identified the challenges associated with assessing military information systems in the context of "new world disorder" missions.

Third, considerable creativity has gone into developing new tools better suited to assessing information systems. Such advances have characterized system dynamics models (e.g., CAPE), agent-based modeling (e.g., Mana), constructive simulations (e.g., JWARS), federates of constructive simulations (e.g., Pegasus), and virtual simulations (e.g., JVB). The realization that no single tool or type of tool can adequately assess information systems has led to the creative orchestration of tools to exploit their strengths and compensate for their individual weaknesses.

Fourth, experiments that provide insights into the potential contribution of information systems to operational effectiveness are now deemed essential. New military information systems are recognized as a stimulus to new doctrine, organizations, training, leadership and education. These activities are the basis for acquiring the data and developing the models that the assessment community requires.

---

[30]See *Joint Vision 2010* and *Joint Vision 2020,* (both available at http://www.dtic. mil/jv2020).

RESIDUAL CHALLENGES: A NEW AGENDA

Despite all these advances, however, the assessment of military information systems is growing more difficult. Future missions will be much more varied than today's, yet there is great interest in an integrated system that would serve as the basis for all mission areas. And the information challenges are intermeshed with the broader transformation of the military. These changes have profound implications for the military's information system assessment community. A new agenda is needed, one with a more comprehensive analytic construct, new assessment capabilities, and a new culture/process for assessment.

A More Comprehensive Analytic Construct. The COBP makes a good start in describing how to assess a military information system. But DoD must extend it to deal with the full range of "new world disorder" missions. This will entail characterizing a broader range of scenarios, formulating new operational concepts, and deriving associated information needs. Consistent with the emerging interest in effects-based operations, the traditional hierarchy of measures of merit must be reevaluated, and new indications of operational success and failure will be needed. Finally, "soft factors," such as "sensemaking,"[31] must be represented better in community assessments.[32]

New Assessment Capabilities. DoD has historically placed great emphasis on the use of a single simulation to support major institutional decisionmaking processes (e.g., the use of TACWAR in the 1997 QDR). This is a Cold War artifact, one that is inadequate for meeting contemporary assessment needs. There is a growing need to assemble a "tool chest" that can be tailored to address the major

---

[31]Sensemaking is a process at the individual, group, organizational, and cultural level that builds on a "deep understanding" of a situation in order to deal with that situation more effectively, through better judgments, decisions, and actions (Dennis K. Leedem, *Final Report of Sensemaking Symposium*, Command & Control Research Program, OASD(C3I), Washington, DC, October 23–25, 2001 [also available at www.dodccrp.org]).

[32]David S. Alberts, John J. Garstka, Richard E. Hayes, and David A. Signori, *Understanding Information Age Warfare*, CCRP Publication Series, August 2001, p. 141.

problems of interest responsively, flexibly, and creatively. This tool chest must include a mix of exploratory tools (e.g., seminar games, influence diagrams, system dynamics models, agent-based models) that are well suited to effects-based analysis and can be used to identify and explore interesting parts of scenario space quickly and efficiently. The tool chest should also include JWARS, DoD's most recent effort to reflect military information systems and processes in a constructive simulation.

However, a tool such as JWARS will have only a supporting role to play as assessments grow narrower and deeper and thus inevitably require new virtual and live models and simulations, particularly to capture the role of the human in complex systems-of-systems. One of the associated challenges facing DoD is the development of consistent, verified databases, data dictionaries, and glossaries to link the components of the tool chest and ensure they are mutually self-consistent. This new tool chest and associated concepts of assessment will require new education and training for the assessment community.

A New Culture and Processes for Assessment. Military information system analysis must embrace a new culture of openness and cooperation that features rigorous peer review, information sharing, and collaboration across traditional organizational boundaries. In the area of transformation, new offices are emerging to stimulate the process—e.g., OSD's Office of Force Transformation and JFCOM's Project Alpha. The military information analysis community has an important role to play in linking these new entities with the traditional organizations that have been pursuing transformation (e.g., the service planning staffs and test and evaluation organizations).

Finally, with the emergence of homeland security and counterterrorism as major mission areas, additional participants must join the assessment process. These include analysts from other federal agencies, such as the new Department of Homeland Security, as well as from regional, state, and local organizations. The military analysis community must recognize that these other participants come to the problem with different frameworks and vocabularies, limited analytic tools, and little or no experience in dealing with classified information. It will take mutual education and training to forge these disparate entities into an integrated community capable of performing creative, insightful analyses of proposed information systems.

Chapter Twelve

THE "DAY AFTER" METHODOLOGY AND NATIONAL SECURITY ANALYSIS

David Mussington

The development of analytic tools to help those making national security policy is driven by the need for usable answers and the urgency of the threats facing the United States. The interaction of these two drivers has produced an array of approaches that favor insights derived from experience with international phenomena. *After all, conducting empirically valid tests of means-ends relationships in international politics is all but impossible, so implicit models must substitute for hard-to-get experimental data.* Most analysts resort to historical comparisons, reasoning via analogy, or conceptualizations that are mathematically rigorous but empirically dubious or even trivial. The "lessons of history" are said to counsel, variously, caution or haste, conservatism or aggression. Appeasement is seen as dangerous, deterrence as infallible but tenuous. Similarities between cases present and past are endlessly discussed with no firm conclusions possible or persuasive—at least as determined through analysis alone.

Decisionmakers are left with concepts and models that necessarily rest on assumptions about the international system, the decision-making and behavioral imperatives of nation-states, and the relationship of military, economic, and political power to the shaping of international outcomes. How can such broad perspectives be converted into something more immediately usable for analyzing complex international phenomena? Abstractions are inevitably gross generalizations of empirical conditions; when adapted to public policy, they are driven by the special needs of decisionmakers for a simplifying schema to understand complex conditions, and by the high level of uncertainty that characterizes political change.

One source of control (in what would otherwise be an unconstrained process) is to exploit the subject matter expertise of issue area specialists as well as rules of thumb presented by diplomatic and military *practitioners* in national security. These rules can form frameworks for understanding complex phenomena. And if these frameworks are made explicit, public policy analysis can subject them to logical and—to some extent—empirical scrutiny sensitive to temporal, technological, and political-economic factors. A setting comporting to a complex real-world international security problem can thus be represented by an abstraction based on what experts hold to be an assessment of a region, technology, or set of relationships. Yet expert-derived frames of reference are not facts, as such, but facts as implied by the conclusions and insights held by fallible human beings. Creating an environment to help this process along is what exercise designers do. Scenario designs, table-top game structures, decisionmaking simulations, and forecasts of future political-military, economic, and technological change—all of these offer tools to meet such analytic requirements. The Day After exercise methodology is one way to elicit structured expertise that channels specialist knowledge into policy dilemmas faced by decisionmakers in the short, medium, and long term.

This chapter describes this approach and shows how it was applied to real-world problems in two projects. It also discusses the value of the Day After methodology, including the special scenario design requirements necessary for successful usage.

THE METHODOLOGY IN BRIEF

A Day After exercise entails a multistage presentation of a hypothetical future, as shown in Figure 12.1. A scenario is derived from contemporary events and policy dilemmas. For design purposes, the scenario time line is divided into a future history and three steps involving actual game play, as follows.

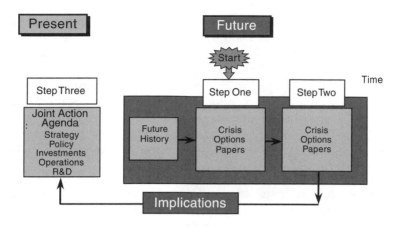

Figure 12.1—Generic Schematic of Day After Exercise Methodology

The Future History. This portion of the Day After exercise methodology is laid out on a time line that starts with the present and then offers a logical sequence of political-military, diplomatic, economic, technological, and policy-related events that identifies the key processes and actors. Background information built into the future history establishes the credibility of the event framework outlined in the next part of the exercise. The more detailed and nuanced the presentation of issues in the future history, the more the follow-on phases can unfold so as to illuminate the policy areas examined.

Step One. Step one involves a policy crisis generated by the actors, processes and entities introduced in a future history that highlights threats to entities or interests important to the United States and/or its allies. The actors are further developed by the decisions they make against a background of specific events. The scenario culminates in a definite escalation. At the end of step one, participants must collectively arrive at decisions appropriate for managing the crisis presented. They must identify the core objectives that policymakers should pursue, and they should decompose the issues entailed in adopting the favored approach to meet those objectives.

Step Two. The crisis escalates steeply in step two in a closely structured and focused evolution of the policy dilemmas and decision imperatives. Although policy responses selected by participants in step one are not directly used in the unfolding events, participants are confronted with step one decisions partially consistent with their step one deliberations. The remaining agenda of responses outlined at the onset of step two intensifies the crisis situation, in order to sharpen the policy dilemmas presented, and challenges the consensus crisis management approach selected earlier by the group. Participants must decide what actions are needed to address the situation effectively, and they must examine the likely consequences of what they decide or fail to decide.

Step Three. In step three, the participants return to the present and are asked to evaluate the situation in light of the exercise experience. Dilemmas are presented from the perspective of contemporary policy choices on the grounds that a framework of prospective actions, policy decisions, and plans could prevent or mitigate the severe conditions described in the scenario narrative. Participants are asked to seek consensus on responses and to clarify areas of pronounced disagreement. Thus, the issue agenda *following* the exercise is addressed so as to advance the identification of potential solutions.

APPLICATIONS AND EXERCISE DEVELOPMENT

The lengthy developmental process responsible for a successful exercise belies the set-piece nature of the Day After approach. Well before the exercise is run, it is tested, different scenarios and issue agendas are explored, and potential participant responses are pondered. An exercise test series helps explore a large number of scenario/issue combinations. The design process subjects this exploration to a disciplined comparative analysis, confronting possible futures with consistent questions and concepts from the perspective of participants in decisionmaking.

Many subjects and issues have been explored using this exercise methodology. Two examples are (1) strategic information warfare (SIW) and mechanisms for addressing significant infrastructure vulnerabilities and (2) electronic commerce technologies (cyberpayments) and international money laundering.

Strategic Information Warfare

Because the sponsor of the SIW exercise was the U.S. Department of Defense (DoD),[1] the focus was on information warfare (IW) as a potential impediment to the exercise of U.S. military options. Those options were themselves predicated on established plans and programs for the timely and efficient delivery of military equipment and personnel to regions designated as strategic to the protection of U.S. friends, allies, and interests. IW threats are introduced directly into existing concepts of strategic security. Although IW is characterized by unique phenomena (i.e., particular weapons and strategic utilization concepts), it is examined in a framework prestructured by well-understood political-military models.

Undertaken in 1995, the SIW Day After exercise was one of the first systematic policy development efforts to explore the potential dilemmas and decisionmaking imperatives associated with society's increasing dependence on information infrastructures. The exercise was developed at a time when basic concepts of vulnerability, crisis stability, and crisis management in the information domain were each relatively unfamiliar. To address this shortfall in rigorous and "high confidence" information warfare conceptualizations, the exercise designers created a hypothetical future in which U.S. critical information infrastructures were targeted by a foreign adversary.

The objectives of the exercise were to

- Describe and frame the concept of strategic information warfare.

- Describe and discuss the key features and related issues that characterize SIW.

- Explore the consequences of these features and issues for U.S. national security as illuminated by the exercises.

---

[1]More precisely, the project sponsor was the Office of the Secretary of Defense (OSD), and the study was defined under the auspices of RAND's National Defense Research Institute.

- Suggest analytic and policy directions for addressing elements of these SIW features and issues.[2]

SIW was framed against challenges—notably asymmetric threats to U.S. national security—deriving from post–Cold War national security imperatives. Information attacks, delivered against information infrastructures accessible via the Internet and the public telephone network, were perceived as potentially serious new vulnerabilities for U.S. military forces.[3]

The SIW exercise used a Persian Gulf scenario, with impacts in the continental United States (CONUS) and Southwest Asia. The Persian Gulf was chosen as the venue for the exercise as the result of a lengthy test series that examined four different scenarios to see which best illuminated policy and strategy issues considered analytically important by RAND researchers and the DoD sponsors. The four scenarios were as follows:[4]

- *Persian Gulf major regional contingency (circa 2000).* Iran seeks hegemony over the Persian Gulf region by overthrowing the Saudi Kingdom via an antiregime organization within Saudi Arabia. A major military crisis develops. The U.S. government decides to deploy forces as a deterrent maneuver. Iran and the local Saudi opposition conduct IW attacks on the Saudi and U.S. governments.

- *Strategic challenge by China in the Far East (circa 2005).* China makes a very aggressive move toward regional dominance. The Taiwanese government declares "independence." China conducts a robust combined-arms military operation, including the use of SIW to deter a forceful U.S. political-military response.

- *Instability in Moscow (circa 1999).* A Russian federation, ruled by a weak central government, is in thrall to several transcontinental criminal organizations (TCOs). A major fissile material diversion

---

[2]Roger C. Molander, Andrew S. Riddile, and Peter A. Wilson, *Strategic Information Warfare: A New Face of War*, MR-661-OSD, RAND, 1996, p. xii.

[3]For a detailed discussion of these issues, see *Critical Foundations: Protecting America's Critical Infrastructures*, the Report of the President's Commission on Critical Infrastructure Protection (PCCIP), Department of Commerce, Washington, DC, 1997.

[4]Molander, Riddile, and Wilson, *Strategic Information Warfare*, p. 6.

to Iran is attempted by a Russian TCO. A Russian TCO makes extensive use of offensive and defensive IW to counter opposition from the United States, several major states within the European Union (EU), and the Russian government.

- *A second Mexican revolution (circa 1998).* The Mexican government faces major challenges from the Chiapas region in southern Mexico and from antiregime movements in northern Mexico. The Mexican revolutionary movements and nongovernmental organization (NGO) allies in North America make extensive use of perception management techniques to dissuade the U.S. government from taking any forceful political, economic, or military action to shore up the beleaguered Mexican regime.

This spectrum of scenarios was adopted to explore in what contexts SIW tools, techniques, and use concepts could be studied. The Persian Gulf was selected to satisfy both the researchers' analytic judgments and the sponsors' policy development requirements,[5] for the following reasons:

- The potential of physically damaging attacks on the United States put in question the physical sanctuary of CONUS.

- A fundamental tenet of U.S. military strategy is to deploy forces to suppress would-be regional hegemons before they succeed and graduate to would-be global hegemons.

- Iran would consider a Persian Gulf scenario to be strategic warfare. Whether a regional adversary uses IW to fracture a coalition or to undermine U.S. or European domestic support for intervention, it plays a strategic game and thus forces the United States into a strategic engagement as well.

- The strategic vulnerabilities and attacks that the United States and its allies might suffer introduce the possibility that other strategic weapons (e.g., nuclear weapons) might be either brandished or used outright.

OSD's planning requirements were a major factor as well. In the end, it is hard to differentiate the scenario's timing and locale from the

---

[5]Molander, Riddile, and Wilson, *Strategic Information Warfare,* p. 8.

analytic reasons underlying scenario selection. This is to be expected where experiential data are obtained using such focused exercise techniques. The policy concepts under examination were a product of independent researcher expertise and repeated interaction with a governmental client. Interactions between clients and sponsors were deliberately structured so that the interests and priorities of the involved departments and senior decisionmakers could be understood. This understanding was then used to write the scenario, which, in turn, was iterated with the sponsor before being subjected to the test series.

Note, however, the interactive nature of the scenario choice process. The exercise tests—which used the other scenarios as well as the one ultimately selected—were each tested in a series lasting from February through June of 1995. Participant feedback on issues, exercise design, and strategy and policy concepts was integrated into each successive test. The entire process was designed to elicit the maximum participant exposure to concepts, scenario variations, and policy problems—in essence, serving as a collective exploration of policy alternatives and possible futures.

Cyberpayments and Money Laundering

The Day After exercise that focused on international money laundering and cyberpayment technologies[6] was undertaken for the U.S. Treasury's Financial Crimes Enforcement Network (FinCEN). The concern was that money launderers would use advanced payment system technologies to conceal their illicit activities.

FinCEN is an interagency clearinghouse for financial crime; it also administers reporting requirements for financial institutions under the Bank Secrecy Act.[7] Many analysts feared that the adoption by narcotics traffickers and transcontinental criminal organizations (TCO) of advanced information technologies would sharply cut the

---

[6]*Cyberpayments* is a term FinCEN coined to describe new payment system technologies that facilitate decentralized (and increasingly, peer-to-peer) value transfers analogous to the exchange of paper currency. Examples of these products include Internet-based e-cash products (such as cybercash) and stored value–type smart card instruments (such as the Visacash and Mondex products).

[7]For information about FinCEN, see http://www.ustreas.gov/fincen/.

efficacy of law enforcement's investigative tools and techniques. Such fears lent urgency to the discussions during the test series. They also opened the research sponsor to new ways of countering emerging patterns of payment system abuse.

RAND analysts, none of whom had any background in researching financial institutions or financial crime, undertook a direct and self-conscious research effort to orient themselves to the money laundering and financial crime landscape. This necessitated a close collaboration with the client and with stakeholders from the financial sector (whom the client helped identify).

The study's objectives were ambitious.[8] The main goal was to explore the dimensions and implications of potential future illicit uses of cyberpayment systems by money launderers and others seeking to conceal funds from governmental authorities so as to identify—at least in a preliminary fashion—possible law enforcement and regulatory responses.

This exercise brought together both public- and private-sector stakeholders. Indeed, FinCEN's interest in the project was driven by hopes that it could serve as a venue for interaction between the public and private sectors. The project's design and question formulation activities were helped by great cooperation from financial firms. As new technologies were anticipated in the payment system, financial industry figures argued for close coordination among state and federal regulatory agencies and private depository institutions. The Day After project served this end and helped to deepen the debate on cyberpayments and the future of payment system technologies.

The research goals of this project were to[9]

• Describe the then-current cyberpayment concepts and systems.

• Identify an initial set of cyberpayment characteristics of particular concern to law enforcement and payment system regulators.

---

[8]See Roger C. Molander, David A. Mussington, and Peter A. Wilson, *Cyberpayments and Money Laundering: Problems and Promise*, MR-965-CTI, RAND, 1998, p. 2.

[9]Molander, Mussington, and Wilson, *Cyberpayments and Money Laundering*, p. 3.

- Identify major issues that cyberpayment policies will need to address.

- Array appropriate approaches to address potential cyberpayment system abuse in a set of potential action plans.

The last goal was particularly challenging. The potential for responses to problems discovered during the exercise was clearly tied to how accurately the future of electronic payment systems was portrayed. The credibility (and longevity) of any recommendations produced by the project were contingent on good technological and market predictions.

Participants included representatives from the executive branch of the U.S. government, the cyberpayment industry, the banking industry, Congress, and academia. Exercise experiences were recorded in the test series underlying the final version of the scenario and in the final operational play itself.

The scenario selected was developed in collaboration with law enforcement officers expert in financial crime and money laundering investigations. It involved a large money laundering system that narco-traffickers used to conceal drug earnings. Funds transfers were concealed by using stored value–type smart cards for street-level drug purchases and then uploading the value into the financial system using merchant stored-value upload terminals. "Participating" store owners received a 4 percent "commission" on each of these transfers. Once the funds reached the financial system, they were electronically transferred offshore using sophisticated layering and integration techniques to hide their ultimate destination.

The scenario narrative then described the compromise of critical technologies used in the manufacture of stored value–type smart cards themselves. This created an even greater threat, one to the integrity of the U.S. and international financial systems. The final portion of the scenario involved a fictional proposal by the Mexican finance ministry to modernize its own banking infrastructure through the adoption of modern electronic banking technologies. This gave money launderers a potential opportunity to penetrate a brand new financial structure and secure unprecedented money laundering capabilities over the long term. The threats to financial system integrity motivated decisionmaking during the future history, step one, and step two.[10]

The exercise findings contained the responses of participants to such dilemmas along with a major analytic examination of those perspectives for integration into competing themes and frameworks. The action plans produced for the report were thus the product of expert assessments by third parties (exercise participants) and post hoc analysis of the deliberations by RAND analysts.

Comparison of the Two Implementations of the Methodology

A comparison of the two projects helps to illustrate the different contexts in which policy analysis interfaces with decisionmaking. In the SIW project, analysts developed scenarios in advance of their real-world appearance by extrapolating known technological trends and factoring in the continuing concerns of the national security community about potential impediments to the execution of U.S. national military strategy. Because U.S. national military strategy focuses on projecting power overseas rather than on defending the homeland, the analysis concentrated on infrastructure vulnerabilities relevant to military preparedness. Similarly, the adoption of asymmetric "counter–information infrastructure" strategies by potential adversaries would place the sanctuary of homeland into question. Hence, infrastructure concerns are related to the concern over the weakening influence of distance as a barrier to potential attacks.[11] These two concerns motivated the selection of both the scenario and the types of attacks presented.

The principal objective of the cyberpayments and money laundering study was to help facilitate a dialogue between government and private-sector personnel on a subject of near-future importance. This project involved a much more in-depth concept-creation process, one in which a community not used to thinking systematically about national strategies was introduced to wide-ranging concepts

---

[10]See Molander, Mussington, and Wilson, *Cyberpayments and Money Laundering,* Appendix B, Exercise Materials.

[11]Homeland security emerged as the preeminent national concern following the events of September 11, 2001.

through the scenario design process. The decentralized nature of law enforcement here gave rise to diverse assessments of long-term trends in criminal activity and to potential investigative counter-measures. Federal law enforcement authorities lead the anti-money laundering arena, but must collaborate with state and local law enforcement officials in individual cases. The international component of anti–money laundering law enforcement activities adds further complexity.

Concepts created in this exercise range from a listed feature set for cyberpayment instruments that defines their importance for potential misuse by money launderers, to a mechanism for tracing Internet-based electronic fund transfers.[12] Both concepts contributed to a framework that defined the technologies of importance to law enforcement and the opportunities for law enforcement to leverage these technologies to enhance investigations. An issue that emerged very quickly during this design process was the lack of clear metrics or measures for anti–money laundering techniques and strategies. Before the project, research sponsors did not appreciate how automation could help evaluate competing strategies for interdicting illicit funds movements. During the test series, this factor was proposed by the design team and emerged as a major focus of future attention for decisionmakers.[13]

Another difference between the two projects stemmed from their differing analyst-sponsor relationships. DoD has had a historically close relationship with RAND (and other independent think tanks). Thus, well-known contracting vehicles and advisory relationships existed for supporting policy analysis. FinCEN, by contrast, had never used independent third-party analysts. The analysts thus needed to foster a close working relationship with FinCEN staff during the exercise design process. Analysts and FinCEN staff had to negotiate who was to author the recommendations of the outbrief report. RAND had to preserve the independence (and peer-review quality control) of the project report's findings while showing sensitivity to the concerns of a client that feared public embarrassment if the exercise results "got out in front of" government policy.

---

[12]Molander, Mussington, and Wilson, *Cyberpayments and Money Laundering*, p. 21.

[13]Molander, Mussington, and Wilson, *Cyberpayments and Money Laundering*, p. 27.

The two exercise programs shared a public sector/private sector character. They served as environments to facilitate dialogue *and* as experimental settings in which policy concepts could be discussed, deconstructed, and critiqued. Although the Day After process hinges on bringing disparate communities of experts together, such interaction must be carefully structured to preserve its analytic independence. In the design phase of a Day After project, new concepts are created and discarded as a way of understanding a subject. Sponsors, however, may interpret such notions as indicators of the project's conclusions. As with sausage-making, the process is messy but the results are often worth the chaos of creative interaction.

The Day After and Analytic Independence

Analysis of the findings is central to the Day After methodology. The exercise designers undertake this analysis, ideally at arm's length from the research sponsor. Because sponsors are extremely engaged in exercise design, they often feel the need to "manage" the production of the report summarizing the project's deliberations and thematic insights. Yet the independence of these two items must be maintained.

A narrative report of findings, records of answers to questions, guidelines used for discussion, and concept creation is, quite appropriately, a shared enterprise; sponsors often provide note-takers and equipment to record deliberations accurately. Nevertheless, analyzing the *meaning* of the facts, insights, and conclusions of experts who interacted within a hypothetical scenario under severe time constraints remains the task of the exercise analysts. They, alone, share a conceptual understanding of the subject matter *and* an architectural knowledge of the scenario construct. Because the exercise scenario necessarily evokes real-world dilemmas, practitioners and policymakers may bring powerful preconceptions to a review of the scenario findings. The Day After method requires that the principal architects of the scenario materials act as filters, differentiating the details and nuances of the story line from the participants' responses to the story. The two projects described above evolved in ways that highlighted the importance of bias control. Both times, RAND was asked to extend the exercise results into a more in-depth examination of the subject matter. The first produced a conceptual framework for understanding the emerging IW policy environment. The second expanded issues addressed in a domestic U.S. setting into an exercise involving 27 countries from the Americas and the Caribbean.[14]

Although details differed in each case, some thematic points can be made. First, the *SIW Rising* document was prepared entirely by the design team, using traditional analytic approaches entailing brainstorming and internal RAND presentations of interim insights, followed by the drafting and redrafting of analytic concepts and models. The resulting policy framework was derived from the reports and exercise experiences achieved during other successful exercise projects. Analyzing those findings generated recurrent themes and insights that contributed to a statement about the nature of the IW setting as it may develop over the next few decades.[15]

Second, the sponsor received the report's analytic framework but not a clear "action plan" for the agency's evaluation. The sponsor already was familiar with the key concepts; many had gained support from Defense Science Board publications and other publicly available research. *SIW Rising* contributed to the policy debate by exposing senior OSD and ASD C3I (Assistant Secretary of Defense for Command, Control, Communications, and Intelligence) staff to concepts that emerged from the non–national security environment that had the potential to affect their plans and priorities.

The follow-on 26-nation cyberpayments study focused on the investigative and prosecutorial implications of money laundering using emerging payment system technologies. It required a new scenario, one that added a section on Internet gambling to the base case of potential misuse of cyberpayments technologies for the purposes of money laundering. The multijurisdictional investigative and prosecutorial nature of the money laundering problem was the focus for much of the design activity, which included employees of the

---

[14]The two reports concerned are Roger C. Molander, Peter A. Wilson, David A. Mussington, and Richard F. Mesic, *Strategic Information Warfare Rising*, MR-964-OSD, RAND, 1998; and David A. Mussington, Peter A. Wilson, and Roger C. Molander, *Exploring Money Laundering Vulnerabilities through Emerging Cyberspace Technologies*, MR-1005-OSTP/FinCEN, RAND, 1998.

[15]Molander, Wilson, Mussington, and Mesic, *Strategic Information Warfare Rising*, pp. 33–38.

Caribbean Financial Action Task Force (CFATF) and the Common-wealth Secretariat of the United Kingdom.

Because the cyberpayments exercise was meant to build a consensus action plan to address legislative, policy, and operational changes in anti–money laundering activities, its scenario had considerable political sensitivity. Thus, the scenario had to be iterated with the research sponsor to a much higher degree than usual, with considerable and detailed dialog taking place on the timing, technical description, and credibility of scenario details. In addition, the test series for the exercise entailed coordinating materials among the 26 countries participating, as well as translating game materials into Spanish.

Interaction with industry representatives provided the required details on Internet gambling and transnational electronic banking trends. Technical details in the exercise were updated with this new information, and elements from the prior exercise were also used. Scenario component reuse is a central feature of Day After scenario design; it leverages issue expertise acquired across a number of different potential projects.

The educational component of the cyberpayments exercise was much more pronounced than any other component (many Caribbean nations lacked basic familiarity with the subject matter). It was important that we educate while avoiding the perception that the Day After approach was "U.S. lecturing." Shared insights and consensus building were the clearest process objectives in the project's execution. The project achieved its objectives, with a draft agenda of priorities resulting from the meeting. In turn, model legislation to respond to many of the policy dilemmas identified within the Day After scenario was to be collaboratively developed by several participating countries. Overall, the exercise helped international dialogue; it also prepared the ground for further policy development.

THE VALUE OF THE DAY AFTER

The Day After method helps decisionmakers and policy analysts address complex subjects in an environment of hypothetical threats to policy goals and objectives. The design of scenarios in this methodology is critical, with plausibility and technical accuracy balanced against the need to focus participant attention on the key themes and analytic issues.

As noted, the Day After requires the creation of a quasi-experiment in which expert insights serve as basic facts in the context of a dynamic scenario. Assumptions that go into the scenario's future history are made as explicit as possible and then honed during repetitive testing of the exercise materials and scenario details.

The Day After methodology can frame futures that challenge analyst and research sponsor assumptions. Can a Day After scenario escape the preconceptions of the designers or the research sponsor? The answer is a qualified yes. It is possible to define scenarios and pose issue questions that challenge the presumptions of both the research sponsors and the researchers. It is not an easy task to accomplish, however, and requires a self-conscious and skilled exercise design approach.

The Day After methodology offers a mechanism for analyzing policy problems. Until now, the methodology has been applied principally to problems with a dynamic component—where technology and technological change affect the policy environment in unpredictable ways. Addressing such situations requires that maximum scope be given to exploring policy futures in order to discern thematic and issue-specific points of decision critical to the goals and/or interests of the research sponsor. This exploration feature of the Day After exercise design process differentiates it from other, more static gaming methodologies.

Lastly, the Day After method educates and raises the consciousness of participants by immersing them in an environment where they suspend disbelief and are made to challenge the views and perceptual frameworks they bring to specific problems. This forces participants to "come up to speed" quickly on complex policy problems and makes them familiar with the characteristically great uncertainty of future-oriented, technologically rich policy environments. The value of this last contribution should not be underestimated. Policymakers face difficult choices, in areas where information is hard to distinguish from advocacy. By offering a critical and rigorous process in which facts and biases are examined, the Day After methodology makes a powerful contribution to the tool kit of policy analysis.

Chapter Thirteen

USING ELECTRONIC MEETING SYSTEMS TO AID DEFENSE DECISIONS

Stuart E. Johnson

Defense decisionmaking is inevitably collaborative because it involves a range of stakeholders. The challenge is to ensure that collaboration adds value instead of producing lowest-common-denominator results.

Collaborative technologies help people develop a common perspective and make it possible to collaborate across time and space. Figure 13.1 is a matrix of collaboration, showing different combinations of time and space—from synchronous (same time, predictable) and colocated (same place) in the upper left, through asynchronous (unpredictable) and uncolocated (different place and unpredictable) in the bottom right. Illustrative collaborative tools and techniques are provided for each space-time combination.

The potential of computer-mediated communications tools for enhancing effectiveness is driving widespread interest in them. Collaboration typically disrupts existing organizational, social, computing, and network infrastructures,[1] so an organizational structure that legitimizes collaboration across hierarchical lines is a key condition of success.[2] Groupware provides a powerful vehicle for transforming

---

[1]See, for example, S. Poltrock, "Some Groupware Challenges Experienced at Boeing," available at http://orgwis.gmd.de/~prinz/cscw96ws/poltrock.html.

[2]See, for example, W. J. Orilkowski, "Learning from Notes: Organizational Issues in Groupware Implementation," *Proceedings of the ACM, Conference on CSCW '92*, 1992, pp. 362–369; B. Vadenbosch and M. M. Ginzberg, "Lotus Notes and Collaboration," *Journal of Management Information Systems*, Vol. 13, No. 3, pp. 65–81; A. S. Clarke

		Time		
		Synchronous (same, predictable)	Asynchronous	
			Predictable	Unpredictable
Space	Same place	Electronic meeting systems	Work shifts	Shared space, group calendaring
	Different places but predictable	Tele/video/ desktop conferencing	Electronic mail	Shared applications and files, collaborative writing
	Different places and unpredictable	Interactive multicast seminars, text chat	Electronic bulletin boards	Discussion databases, workflow systems

Figure 13.1—Collaboration Across Time and Space

stovepipe processes into more-integrated decisionmaking. Defense planning, in particular, can be dramatically improved.

This chapter addresses how one kind of groupware, electronic meeting systems (EMSs), can be used for simultaneous collaboration. Also included is a description of the detailed application of one EMS (Ventana Corporation's GroupSystems) to defense planning.

ELECTRONIC MEETING SYSTEMS

An EMS includes three processes that are designed to improve group productivity.[3] Table 13.1 shows these processes and some of their advantages.

(ed.), *Groupware: Collaborative Strategies for Corporate LAN's and Intranets,* Prentice Hall, Upper Saddle River, NJ, 1997; and http://copernicus.bbn.com/lab/ocsc/papers/ Full.text.html.

[3]For more detail, see Jay F. Nunamaker, Jr., Robert O. Briggs, and Daniel D. Mittleman, "Electronic Meeting Systems," in David Coleman and Raman Khanna (eds.), *Groupware: Technology and Applications,* Prentice-Hall, Inc., 1995.

Table 13.1

EMS Processes and Some of Their Advantages

Process	EMS Advantages
Communication through common media (computer network, videoconference, teleconference)	Increases the number of people who can participate in a meeting through simultaneous input
Thought processes to form an action plan to accomplish a common goal (formulate, evaluate, and select or prioritize alternatives)	Generates more ideas of higher quality through various collaborative activities to generate, organize, and evaluate ideas; anonymity allows free debate on ideas
Information access to enable group members to support the thought processes using timely, accurate, and complete information	Tools can reduce information overload, increase productivity through access to a larger information base, and enhance organizational learning via electronic transcripts

An EMS focuses on group dynamics, using computer-aided parallel communications, structured and focused thought processes, and applications and tools to improve information access. Used well, an EMS can enhance defense planning by stimulating social interaction and thinking, and can accelerate strategic planning, problem solving, and the setting of priorities because its ability to exploit simultaneous input facilitates idea generation, persuasion, and decision selection.

The results produced by a decision support process depend on the participants, the leadership, and the exercise design, as well as on the underlying technology. Successful decision support exercises are those that identify and represent the problem clearly, generate and evaluate alternatives, and then select among those alternatives. An idealized flow chart of the process shows its phases:

- Define the problem
- Formulate a decision objective
- Generate decision criteria, weighting them as appropriate
- Generate alternatives
- Discuss alternatives

- Prioritize alternatives

- Rate how well each alternative meets each criterion

- Compare the scores for the alternatives and prioritize the alternatives accordingly

- Capture the pros and cons of each leading alternative for presentation to decisionmakers.

PRIORITIZING NAVAL PROGRAMS: AN EXAMPLE OF AN EMS IN USE

The Challenge

In early 1997, the Navy staff (N-8) was tasked to prepare the Navy for the first Quadrennial Defense Review (QDR). The Navy's dilemma was familiar to force planners: the Navy had developed a program in response to the defense planning guidance, and that program exceeded the *fiscal* guidance laid down by the defense secretary. Navy leadership argued to the Office of the Secretary of Defense (OSD) that it had long since eliminated the fat from its budget and requested additional funding. Simultaneously, the Navy moved to ensure that if it could *not* get additional funding, it would have the best backup plan it could devise.

In formulating its backup plan, the Navy asked the Decision Support Department (DSD) of the Naval War College to develop a methodology that would force planners to do collaborative, capabilities–based planning. DSD responded with an analytic exercise (carried out in spring 1997) that drew on decision support technologies and techniques. The results were delivered to the N-8 staff for incorporation in the Navy's input to the QDR.

The Navy's challenge was how to fit a $90 billion requirement into a fiscal guidance of $81 billion.[4] One time-honored approach would have each head of a major program element resubmit a budget that trims that program by 10 percent. This may well be the "easiest" approach in that it introduces the least stress across the organization. It

---

[4]All budget quantities are expressed in fiscal year (FY) 1997 dollars.

seems "fair," because everyone takes the same hit. But it almost always causes serious disruption from an overall planning perspective. Some programs can take a 10 percent cut with only marginal reductions in their effectiveness; others are so crippled by the cut that they might as well be eliminated.

Another common approach is to lock the heads of major programs in a room and don't let them out until they have come up with a plan that fits the budget. The problem with this approach is clear: *Program prioritization becomes subject to competition—and then compromise—among the major program sponsors.* The distortion in the resulting program is evident when one looks at the uneven capabilities that result from such bargaining.

The alternative that the Navy leadership needed was clear: *a process to prioritize programs based on capabilities that naval forces need.* This is easier said than done, however. Many capabilities go into a military force, and they can be delivered in many ways. Targets, for instance, can be struck from the sea by carrier aircraft or missiles; missiles, in turn, can be launched from a surface ship or a submarine. The DSD staff set out to design an exercise that would

- Tie program priorities to required capabilities
- Involve hands-on participation by key resource sponsors
- Be transparent to participants
- Provide a clear audit trail of results.

At the outset, the exercise design team faced the fact that program prioritization is always dangerous. Every program that has made it far enough to be in the Navy's program objectives memorandum (POM) is important; all of the programs bring some important military capability to the table. The challenge that Navy leadership laid down was as follows: programs that were multipurpose and high performing or narrow-purpose but critical to fielding key capabilities were to be identified and distinguished from programs that were narrow-purpose and whose contribution was marginal or could be covered in another way.

DSD was asked to concentrate on the investment account. True, the Navy would also look elsewhere to save money; its operations and maintenance costs refused to decline even though every year saw fewer ships in the fleet. Additional rationalization of the Navy's base structure and facilities could also save money. Nevertheless, Navy leadership felt it had a good understanding of *where* to find savings in these other accounts; it simply did not know *how* to do so. The investment account was different, however. Developing and procuring one system when another would serve much better could haunt the Navy for as long as 30 years.

Overview of Methodology

The analytic approach adapted the strategy-to-task methodology developed at RAND during the 1980s and married it to decision support technology and methods. Experts who understood U.S. national security strategy as articulated in the President's National Security Strategy and the secretary of defense's guidance to the armed services were assembled. Their understanding of this and their knowledge of activities the military had been called on to perform in the recent past would enable them to project the activities military forces would likely be asked to carry out within the foreseeable future. This, in turn, would help them judge what capabilities the Navy would need. From there, it would be a straightforward, if complex, step to determining the programs needed to field those capabilities.

The design of phase I of the exercise—through determining capabilities—is shown as a flow chart in Figure 13.2.

Once a weighted set of capability requirements has been established, the next step is to assess how well the Navy programs satisfy those weighted requirements. Figure 13.3 shows the final output as a prioritized list of major programs in the Navy's investment budget based on the contribution each makes to the weighted set of capabilities the Navy would need. The exercise flow, and the EMS's part in it, can then be described in detail.

Phase I: From Activities That Military Forces May Be Called on to Carry Out to Weighted Capability Requirements

Select the set of activities that military forces may be called on to carry out. The U.S. military responds to tasking from the national command authority that ultimately culminates in the authority of the

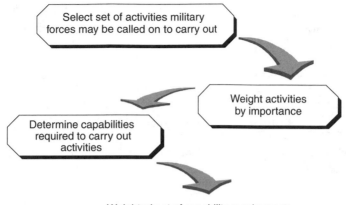

Figure 13.2—Analytic Approach, Phase I

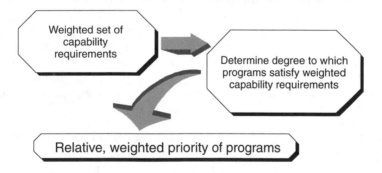

Figure 13.3—Analytic Approach, Phase II

U.S. president as commander in chief, with supplemental "advise and consent" authority resting in the U.S. Congress. Some 20 analysts (both military and civilian) from the Naval War College who were familiar with the details of the national security decisionmaking process gathered to develop a list of activities that military forces could be called on to do over the next two decades. Each analyst was asked to enter into the EMS notebook computer specific suggestions of activities with short descriptions. Each activity was displayed both on each participant's notebook computer and on a screen in the front of the room that was seen by all participants. The ability to see all the inputs limited the duplication of activities nominated and stimulated thinking.

When the participants had entered all the "activities" they could think of, the meeting facilitator led the group to review candidate activities displayed on the master screen. The group collectively examined the list, removed redundancy, and made the language more precise. The result was a long list of activities (listed in Table 13.2) that these experts felt the military could be called on to do in the foreseeable future.

A long unprioritized list is a poor foundation for making the hard choices, tradeoffs, and risk assessments that defense planners face in every Planning, Programming, and Budgeting System (PPBS) cycle. So the facilitator turned to the difficult task of prioritization.

Weight activities by importance. In the Bottom-up Review of 1993, Secretary of Defense Les Aspin had made a clear and simple judgment on the question of priorities: *The US military was to prepare to fight and win two nearly simultaneous major regional contingencies.* If the military met this standard, it could cope with other, "lesser included" contingencies. Hence, the guidance was, essentially, to prepare for the worst, and the resultant capabilities would be equal to any other challenge.

The group of experts felt that this formulation, while useful, was an oversimplification and broke down on careful examination. The devil was in the details. Some activities—e.g., maintaining a peace accord in Bosnia—turned out to be "different" cases, not "lesser included" cases. The equipment, types of units needed, and skills required were different from those of a force optimized for high-intensity conflict. Moreover, how much special attention to pay to "other" activities would have to depend on how important they were, which, the group decided, was determined both by the likelihood that the military would be called on to carry them out and the risk to U.S. national security interests were they not carried out successfully.

Militaries are constantly undertaking activities quite different from fighting a major conflict, such as maintaining a forward presence to deter aggression or monitoring international agreements. And then there are some activities—such as nuclear warfare—that are highly

Table 13.2

Missions for DoD over the Next 20 Years

Short Form	Long Form
Constabulary	Provide constabulary assistance to U.S. domestic authorities
Counterdrug	Assist U.S. civilian agencies in countering drug trafficking
Counterimmigration	Help U.S. civilian agencies interdict illegal immigration
Counterinsurgency	Protect democracies by conducting counterinsurgency
Counterproliferation	Actively support counterproliferation activities
Counterterrorism	Assist U.S. civilian agencies in countering terrorism
Crisis response	Respond to a crisis rapidly
Deter MTW	Deter a major theater war (MTW)
Deter war with peer	Deter major war with a peer competitor
Deter WMD	Deter development and use of weapons of mass destruction (WMD)
Extend deterrence	Extend deterrence and defense coverage to a friendly nation
Fight and win MTW	Fight and win a MTW
Forward engagement	Conduct forward engagement
Humanitarian ops	Conduct humanitarian relief operations
Impose U.S. will	Impose U.S. will through military intervention
Intelligence	Collect intelligence
Int'l agreements	Monitor and enforce international agreements
Limited ops	Conduct limited operations to influence a major power
Peace ops	Support and/or conduct peacekeeping operations
Protect U.S. lives	Protect U.S. lives and property (to include noncombatant evacuation operations)
Punitive strikes	Conduct limited punitive strikes
Sanctions	Enforce sanctions

unlikely but would carry serious consequences for national security if they were to happen and the United States were unprepared. Both likelihood and consequence are thus integral to assessing an activity's importance. To capture these two factors, the facilitator asked participants to assess activities against two assertions:

1. The military is likely to be called on to carry out this activity.

2. Not performing this activity successfully poses significant risks to U.S. national security.

The participants assessed each activity, one at a time, against each assertion, selecting responses ranging from strongly disagree (1) to strongly agree (5). The EMS alternative analysis module then displayed the mean, standard deviation, and range of responses for each military activity. The responses entered by participants into their notebook workstations were anonymous (although participants could compare their responses against what the group did as a whole) in order to eliminate any influence some participants might feel from others with higher rank or stronger personalities. Truly independent assessments thus were possible. Participants were encouraged to append explanations to their assessments, which, in turn, would be displayed. This yielded a much richer understanding of the results, especially any "outliers."

The responses were then binned. Activities with mean scores of 4.5 or above were placed in the Strongly Agree category, those with mean scores of 3.5 to 4.4 in the Agree category, and so on. The results are shown in Figures 13.4 and 13.5.

The facilitator then paused to review these results with the participants. Had the ensuing discussion revealed any misunderstandings of the assertions, the facilitator could have "polled" the group again once the assertions had been clarified. Repeating the poll was just a mouse-click away.

The group felt that frequency and risk were of comparable importance in evaluating activities and concluded that they should be given equal weight in a consolidated ranking (see Figure 13.6).

Note that the emerging numerical priority ranking constituted an important step beyond many strategy documents, which catalog a broad list of military missions but then implicitly concede that the next step, prioritization, is too hard, and so stop.

The group paused to reexamine whether giving equal weight to likelihood and consequence, or "criticality," was appropriate. One participant argued that because the military's job was to protect the United States and its interests from catastrophic harm, the ability to do so was the standard by which force capabilities should be judged.

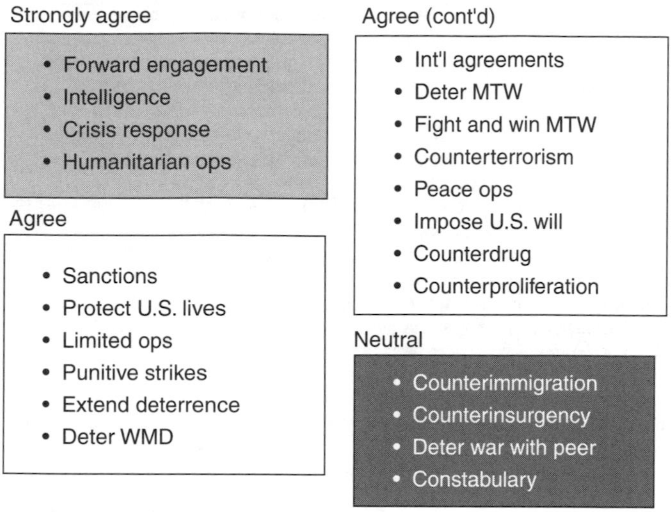

Strongly agree

- Forward engagement
- Intelligence
- Crisis response
- Humanitarian ops

Agree

- Sanctions
- Protect U.S. lives
- Limited ops
- Punitive strikes
- Extend deterrence
- Deter WMD

Agree (cont'd)

- Int'l agreements
- Deter MTW
- Fight and win MTW
- Counterterrorism
- Peace ops
- Impose U.S. will
- Counterdrug
- Counterproliferation

Neutral

- Counterimmigration
- Counterinsurgency
- Deter war with peer
- Constabulary

Figure 13.4—Responses to "The Military Is Likely to Be Called on
to Carry Out This Activity"

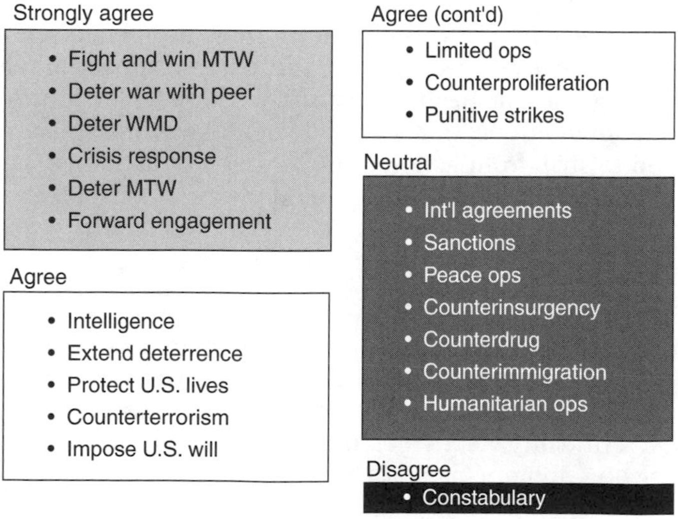

Strongly agree

- Fight and win MTW
- Deter war with peer
- Deter WMD
- Crisis response
- Deter MTW
- Forward engagement

Agree

- Intelligence
- Extend deterrence
- Protect U.S. lives
- Counterterrorism
- Impose U.S. will

Agree (cont'd)

- Limited ops
- Counterproliferation
- Punitive strikes

Neutral

- Int'l agreements
- Sanctions
- Peace ops
- Counterinsurgency
- Counterdrug
- Counterimmigration
- Humanitarian ops

Disagree

- Constabulary

Figure 13.5—Responses to "Not Performing This Activity Successfully
Poses Significant Risks to U.S. National Security"

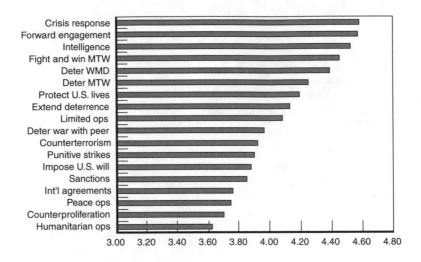

Figure 13.6—Ranking of the Importance of Military Activities

Others countered that U.S. military forces were being called on to respond to any number of situations that posed no serious threat in the short run to U.S. national security but that nonetheless required a competent response for two important reasons. First, a modest size crisis, if not checked promptly, could expand into a much larger problem. Second, the need to respond to these lesser contingencies was not going to disappear. To meet such needs, the military would be forced to strip from its forces units that would otherwise be earmarked as lead elements for any MTW that might break out. Planning deliberately for activities that must be done anyway would thus allow the military to minimize turbulence in its forces.

In this case, an equal weighting was retained. Had it not been, the EMS software would have permitted sensitivity analyses to be performed so that the group could see how priorities would change if the weighting changed. Thus, if the weighting moved to, say, 3:2, in favor of "criticality," activities that implied high-intensity warfare with a strong military (such as "deter war with a peer") would have increased modestly in importance.

Rank the relevance of naval forces in carrying out key military activities. The facilitator then asked the group to consider what naval forces could contribute to joint operations that would increase the odds of those operations succeeding. Naval forces would play an important role in most of the activities, but in some their role would be more prominent, and in others they would be less suited than the forces of another service. In some cases, naval forces would have the primary responsibility; in others, a critical role (i.e., one without which the operation would be severely hampered); in yet others, only a marginal role.

The facilitator asked participants to assess activities against two assertions:

1. This activity is likely to be performed primarily by naval forces.

2. Naval forces are critical to the performance of this activity.

After discussing the role naval forces would play in carrying out these activities, the participants entered responses into their notebook workstations. One by one, each mission was displayed, and each participant entered a response from strongly disagree (1) to strongly agree (5). Figures 13.7 and 13.8 show the results after aggregation and binning. As expected, the two responses are highly correlated, as Figure 13.9 shows.

The results of these two polls were then averaged to portray the overall "utility" of naval forces for carrying out activities the military would be called on to execute. The top 15 priorities are shown in Figure 13.10.

Determine the capabilities that naval forces will need. The prioritized list (Figure 13.10) gave the group a basis for identifying the capabilities needed and, ultimately, which programs would provide those capabilities.

As part of the 1990s effort to build a Joint Mission Essential Task List, the Navy developed its Department of the Navy (DoN) Warfare Task List, which sets out the capabilities that the Navy commits itself to maintaining in its forces. This list is reviewed and updated regularly to ensure it takes into account new capabilities that technology or new doctrine make possible. It is arranged hierarchically, which means the major categories lent themselves well to the exercise at

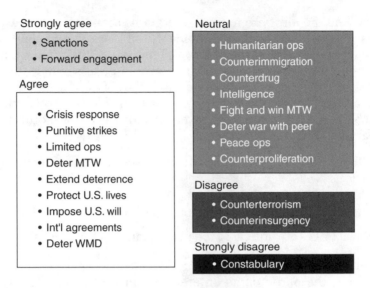

**Figure 13.7—Responses to "This Activity Is Likely to Be Performed
Primarily by Naval Forces"**

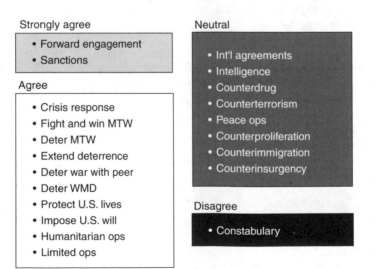

**Figure 13.8—Responses to "Naval Forces Are Critical to the Performance
of This Activity"**

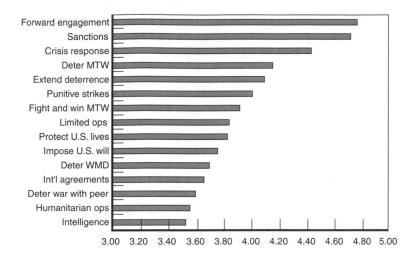

Figure 13.9—Ranking of the Relevance of Naval Forces in Carrying Out Key Military Activities

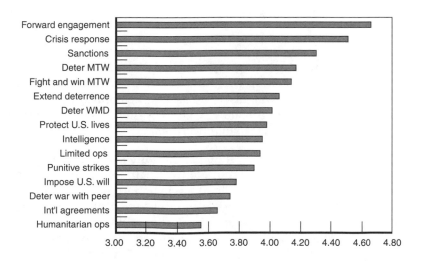

Figure 13.10—Ranking of the "Utility" of Naval Forces

hand. The DSD team preselected the following DoN warfare tasks from the list to use in arraying the various naval capabilities required:

- Airspace primacy

- Command and control of assigned U.S. and multinational forces

- Control of littoral land areas

- Fire support to forces ashore

- Forcible amphibious entry

- Forward deployed, combat-capable forces

- Gather and disseminate timely battlespace knowledge

- Precision strike

- Sub-surface primacy

- Surface primacy

- Sustained expeditionary logistics

- Theater missile defense

- Timely movement of forces and supplies by sea

One by one, each activity (e.g., fight an MTW) was displayed on the participants' screens, along with, one by one, the capabilities to be evaluated. Participants were asked to assess the contribution the capability shown made to the activity shown, the possible responses ranging from not critical (1) to highly critical (5).

The "score" for each activity-capability combination was the product of the activity's weight (shown in Figure 13.10) and the capability's degree of criticality. For example, if a participant deemed that providing theater missile defense was a capability critical (4) to the crisis response activity (4.5), the "score" for that combination would be 18 (i.e., 4 x 4.5). These scores were then summed for each capability to produce a "weighted capability requirement." A capability scored "high" on the list of weighted capabilities if it contributed disproportionately to the activities with the highest weights, or if it contributed a high value to a broad spectrum of missions. The results for the top 13 scorers are shown in Figure 13.11.

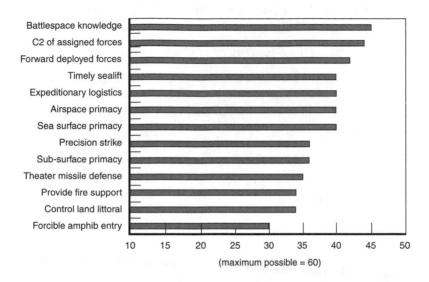

Figure 13.11—Capabilities Ordered According to Their Contribution to Weighted Activities Naval Forces Are Likely to Be Called on to Carry Out

Of particular note is the lead priority given to battlespace knowledge and effective C2. *Neither* of these two "most important" capabilities puts firepower or forces into the field. Instead, both allow the commander to understand the situation in the battlespace and to apply force or firepower more precisely and effectively.

The group singled this out as an important finding made possible by the methodology. Normally, discussions of capabilities measure effectiveness by focusing on the amount of force or, in more refined analyses, the amount of firepower a service can field. This focus biases programmatic priorities toward major weapons platforms. Because the value of battlespace knowledge and effective C2 is hard to measure and is rarely part of high-visibility programs, these two capabilities are typically put at a disadvantage when competing for resources. Even careful modeling and simulation generally understate their value. By now, most models can elevate the measure of merit to targets killed, thereby picking up the importance of such items as precision guided munitions (PGMs). But battlespace knowledge and effective C2 remain elusive capabilities to model, so they seldom are assessed in ways that give prominence to their payoff.

Phase II: Assessing the Department of the Navy Investment Program

As the last phase of the exercise, the group turned to the final task: prioritizing the Navy's investment program by assessing the largest DoN programs.[5] These 22 programs, which represented 92 percent of the total DoN investment budget, were as follows:

- New class of aircraft carriers
- F-18 E/F combat aircraft
- V-22 Osprey
- DDG-51 (with Aegis air defense system)
- Air-cushioned amphibious vessel (ACAV)
- Joint Strike Fighter (JSF)
- Lightweight 155-mm artillery piece
- Arsenal ship
- SC-21 (now the DD21) destroyer
- AV-8B V/STOL remanufacture
- F-14 upgrade
- Transport helicopter upgrades
- P-3 Orion upgrade
- Standoff precision guided machines (e.g., JSOW, JDAM)
- AIM 9X air-to-air missile
- LPD 17 amphibious transport
- Theater ballistic missile defense (TBMD)
- Surface fire support (NSFS)

---

[5]Based on the sum of procurement and 6.3 (research and development [R&D]) dollars across the 1997–2002 FYDP.

- New attack submarine (NSSN)

- Unmanned aerial vehicles (UAVs)

- Cooperative engagement capability (CEC)

- EA-6B electronic warfare aircraft

Participants were presented with a description of each program and its place in the budget. They were then asked to evaluate the contribution—from not critical (1) to highly critical (5)—that each program made to the weighted capabilities derived in phase I. As the capabilities were displayed, one by one, participants assessed the contribution the program would make to them.

The "score" assigned to each capability-program combination was the product of the capability's weight and the participant's assessment of the degree to which the program contributed to the capability. These scores were then summed for each program to produce a weighted priority list of programs in the DoN budget. Figure 13.12 shows the results.

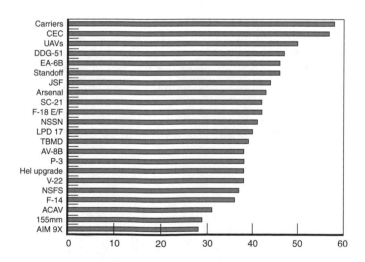

Figure 13.12—Weighted (Joint) Priority of Largest Navy Programs

A program could score high on the list of priorities if it contributed disproportionately to the capabilities with the highest weights, if it contributed at least moderately to a broad spectrum of capabilities, or if it contributed a high value to enough capabilities to push its score up. In this way, systems that contributed to a set of high-value capabilities or were important to a broad set of capabilities rose to the top.

A review of the results reveals the power of this analytic tool. The program for future carriers scored first. To be sure, it is hard to imagine a group of naval officers ranking the next generation of aircraft carriers anywhere other than first. That said, however, a carrier is a highly flexible military system, one that contributes a broad range of the required capabilities to the (weighted) military activities identified by the group.

More surprising to the naval leadership was the high priority that emerged for both CEC and UAVs. An examination of participants' entries, where they "explained" their assessments, revealed that this high priority grew out of a perception that these were critical to the high-priority capabilities they had identified earlier (see Figure 13.11). This, in turn, grew from the assessment that future military operations are likely to be carried out in an environment characterized by enemies with anti-access capabilities that include cruise and ballistic missiles. A good picture of the battlespace was judged to be critical, as well as a measure of defense—hence, the high priority assigned to the DDG-51 with its advanced air defense radar and potential to develop an anti–ballistic missile and anti–cruise missile capability. The same anti-access threat drove the robust standoff munitions buy to a high priority as well.

Other Possible Uses

This process yielded a list of programs prioritized by their contribution to the capabilities the Navy will need in the future and to the activities the Navy will likely be called on to undertake. The process provided a systematic look at the relative contribution each competing program (none of which was unimportant) might make to the naval service's future. The resulting list allowed senior Navy decisionmakers to focus on programs that ought to be fenced off as far as possible in budget deliberations. Yet this EMS technique is by no means limited to determining priorities for capabilities or programs. It could be applied to operational concepts, R&D priorities, or any decisionmaking process being conducted under conditions of uncertainty.

This book reflects the reshaping of defense analysis that went on at RAND and elsewhere after Desert Storm and the fall of communism—i.e., the movement away from a focus on the awesome predictability of the Soviet threat toward contemplation of a world that had become, if less dangerous, surely more uncertain. In place of the Cold War's one overriding potential foe, there were now numerous possible threats, so the Department of Defense (DoD) moved, haltingly, from single planning scenarios to planning for two major wars and, ultimately, to planning based not on threats or scenarios but on the capabilities needed for a fast-changing world. It is no surprise that this book, and the body of work that lies behind it, is dominated by techniques for dealing with uncertainty and, uncertainty's obverse, the explosion in information that also characterizes today's world—and tomorrow's.

In its efforts to understand and frame policy for this changing world, RAND developed new analytic techniques, such as exploratory modeling, and a wider range of its analyses became relevant, reflecting the changing nature of security. Peace and humanitarian operations, for instance, required the U.S. military to deal with not only a wide range of coalition partners, but also with private nongovernmental organizations (NGOs). The same was the case as RAND thought through the problem of protecting critical national infrastructures (such as that for information), infrastructures that are no less a national asset for being mostly in private hands.

The terrorist attacks of September 11, 2001, have now ushered in a new world, one still far from being fully apprehended by the American people or their leaders. Not only have the potential threats become still more diverse, but some of the fundamental distinctions of planning have been overturned. The United States planned—and still must plan—to project force over large distances. But it now must also deal with threats right at home. Comfortable distinctions between "foreign" and "domestic" were already being eroded by the rise of the global economy. Terrorism does not respect them at all.

The world before us will require RAND and its fellow organizations to further develop the approaches and techniques in this book. Planning will be increasingly stretched by the blizzard of uncertainty and the range of capabilities that might be needed. How much and how the U.S. military will be involved in homeland security remain to be seen, but surely the change from the world of the Cold War will be marked. Interactions defying other comfortable distinctions—those between "government" and "the private sector"—will also intensify, calling forth new ways of understanding new kinds of partnerships.

Terrorism surely is a grave threat, but it is not exclusively or even primarily a *military* threat—another manifestation of the changing nature of security. Now, as the world confounds comfortable distinctions, it also calls on a wider range of RAND analyses and impels collaborations that were infrequent before. Techniques from what used to be thought of as RAND's "domestic work" are more and more relevant to the nation's security, rather than just its well-being. RAND Health, for instance, has moved quickly to make its knowledge of medicine and health-care delivery available to those planning against biological or chemical terror. The kind of partnership between health care and national security professionals that was often discussed but seldom occurred is now happening. RAND's next volume on challenges and techniques for defense and security decisionmaking will be enriched by work from not just health, but from survey research, criminal justice and public safety, insurance and infrastructure protection, and other realms of RAND research.

Bruce W. Bennett (Ph.D., policy analysis, RAND Graduate Institute for Public Policy Analysis) is a senior analyst and professor of policy analysis at RAND. His research interests include military strategy and force planning, countering proliferation of weapons of mass destruction, Korea, and the Persian Gulf.

Nurith Berstein (Masters in public administration, Carleton University, Ottawa, Canada) is a researcher at RAND specializing in national and international security policy issues.

Frank Camm (Ph.D., economics, University of Chicago) leads research at RAND on high-level Army resource management issues associated with force structure design, logistics policy, and acquisition of combat service support services.

David S.C. Chu (Ph.D., economics, Yale University) has served in a variety of RAND and national security government posts.

Paul K. Davis (Ph.D., chemical physics, Massachusetts Institute of Technology) is a senior scientist and research leader at RAND. His research encompasses a number of areas, including strategic defense planning, future forces and force transformation, and advanced modeling and simulation. He is a professor at the RAND Graduate School; a member or former member of the Naval Studies Board of the National Research Council, the Defense Science Board, and the U.S. SALT Delegation; and author of numerous books and studies.

Daniel B. Fox (Ph.D., operations research, University of Illinois) is a senior operations research analyst at RAND. His focus is the design and application of computer simulations for analysis of complex military operations. Dr. Fox has over 30 years of experience in military operations analysis.

James R. Hosek (Ph.D., economics, University of Chicago) is a senior economist at RAND, editor-in-chief of the *RAND Journal of Economics*, RAND Graduate School professor, and former director of RAND's Defense Manpower Research Center.

Stuart E. Johnson (Ph.D., physical chemistry, Massachusetts Institute of Technology) was a senior scientist at RAND when this research was completed. He has many years of experience in defense planning and analysis at the Department of Defense, NATO headquarters, and in private industry.

Martin C. Libicki (Ph.D., city and regional planning, University of California, Berkeley) is a senior policy analyst at RAND whose areas of expertise include the application of information technology to national security.

David Mussington (Ph.D., political science, Carleton University) is a political scientist at RAND. Among his areas of expertise are critical infrastructure protection, information and technology security, and counterterrorism and cyberterrorism.

Stuart H. Starr (Ph.D., electrical engineering, University of Illinois) is the Director of Plans at The MITRE Corporation; his areas of expertise include the assessment of information systems in the context of national security missions.

Harry J. Thie (Doctorate, business administration, George Washington University) is a senior management scientist at RAND whose research explores military career management and defense organization, manpower, personnel, and training.

Gregory F. Treverton (Ph.D., economics and politics, Harvard University) is a senior policy analyst at RAND. He has worked on intelligence and on Europe for Congress, the White House and the National Intelligence Council; his current research interests also include Asia and public-private partnerships.

DATE DUE

JUN 2 6 2007

7/9/09